GREAT MOMENTS
IN
MATHEMATICS
(AFTER 1650)

By
HOWARD EVES

THE
DOLCIANI MATHEMATICAL EXPOSITIONS

Published by
THE MATHEMATICAL ASSOCIATION OF AMERICA

———

The Dolciani Mathematical Expositions

NUMBER SEVEN

GREAT MOMENTS
IN
MATHEMATICS
(AFTER 1650)

By
HOWARD EVES
University of Maine

Published and distributed by
THE MATHEMATICAL ASSOCIATION OF AMERICA

The DOLCIANI MATHEMATICAL EXPOSITIONS series of the Mathematical Association of America was established through a generous gift to the Association from Mary P. Dolciani, Professor of Mathematics at Hunter College of the City University of New York. In making the gift, Professor Dolciani, herself an exceptionally talented and successful expositor of mathematics, had the purpose of furthering the ideal of excellence in mathematical exposition.

The Association, for its part, was delighted to accept the gracious gesture initiating the revolving fund for this series from one who has served the Association with distinction, both as a member of the Committee on Publications and as a member of the Board of Governors. It was with genuine pleasure that the Board chose to name the series in her honor.

The books in the series are selected for their lucid expository style and stimulating mathematical content. Typically, they contain an ample supply of exercises, many with accompanying solutions. They are intended to be sufficiently elementary for the undergraduate and even the mathematically inclined high-school student to understand and enjoy, but also to be interesting and sometimes challenging to the more advanced mathematician.

———

The following DOLCIANI MATHEMATICAL EXPOSITIONS have been published.

Volume 1: MATHEMATICAL GEMS, by Ross Honsberger

Volume 2: MATHEMATICAL GEMS II, by Ross Honsberger

Volume 3: MATHEMATICAL MORSELS, by Ross Honsberger

Volume 4: MATHEMATICAL PLUMS, edited by Ross Honsberger

Volume 5: GREAT MOMENTS IN MATHEMATICS (BEFORE 1650), by Howard Eves

Volume 6: MAXIMA AND MINIMA WITHOUT CALCULUS, by Ivan Niven

Volume 7: GREAT MOMENTS IN MATHEMATICS (AFTER 1650), by Howard Eves

A Grateful Acknowledgment

I wish to express my sincerest thanks to Professor Ross Honsberger and Professors G. L. Alexanderson, J. Malkevitch, and K. R. Rebman, Chairman and Members of the Dolciani Subcommittee, for their careful and constructive reading of the original manuscript. Their criticisms and suggestions were invaluable.

HOWARD EVES

Dedicated to Carroll V. Newsom,
in friendship and gratitude

PREFACE

In this second volume of GREAT MOMENTS IN MATHEMATICS we continue the lecture sequence started in the first volume. Having, in the first volume, presented twenty lectures devoted to outstanding events in mathematics that occurred prior to 1650, in this second volume we devote twenty lectures to outstanding events in mathematics that occurred after 1650. Here the choice of topics was much more difficult than for the first volume; the modern era in mathematics is so rich in fine candidates for GREAT MOMENTS IN MATHEMATICS that, for sheer lack of space, many had regrettably to be passed over.

The reader will find, as is naturally to be expected, that this second volume is somewhat more demanding than the first volume, but a reasonable acquaintance with beginning differential and integral calculus should suffice for a proper understanding of the material.

Once again, we must apologize for the loss of sparkle and excitement suffered when oral lectures are presented in abbreviated written form. The multitudinous array of lecture props (overhead transparencies, colored slides, maps, portraits, objects, desk experiments, etc.) must be forsaken, and, for lack of space, many charming anecdotes and related stories must be omitted. How much less impelling and stimulating, for example, is the written presentation of LECTURE 29, devoted to the thrilling discovery of the first noncommutative algebra. Gone are the transparency portraits of Hamilton, Grassmann, Cayley, and Gibbs; the set of colored slides of a pilgrimage to the Brougham Bridge over the Royal Canal near Dublin, with a shot of the plaque commemorating Hamilton's sudden realization on the evening of October 16, 1843, at that very spot, of the quaternionic multiplication table, and shots of the famous mathematician's favorite walk along the canal tow-path; pictures of Trinity College of Dublin, the Dunsink Observatory, specimens of Hamilton's poetry, and other Hamiltonian memorabilia; and the insightful anecdotes about Ireland's foremost mathematician.

HOWARD EVES

Fox Hollow, Lubec, Maine
Winter 1978–79

CONTENTS

Allison →

Traci →

Renee

ORDER WITHIN DISORDER

Among the bones of the intricate skeletal structure of the foot is one lying in the heel just above the talus bone and known as the *astragalus*. In man, and in animals with a developed foot, the astragalus is quite irregular, but in the hoofed animals, like sheep, goats, and all kinds of deer, the astragalus has a rough symmetry, being squarish in cross section with two rounded ends, one slightly convex and the other slightly concave. These bones are solid and essentially marrowless, hard and durable, somewhat cubical with edges measuring an inch or less, and, with handling, capable of taking on a high polish.

It is not uncommon for archaeologists excavating at prehistoric sites to find sizable collections of astragalus bones of small hoofed animals, and sometimes collections of small stones of various colors. It seems reasonable to conjecture that these bones and pebbles may have been used by prehistoric man as tally-stones or counters, and as toys for himself and his children. While such a use for astragali in prehistoric times is only conjectural, there is no doubt that among the ancient Babylonians and Egyptians, and the Greeks and Romans of the pre-Christian era, one of the uses of astragali was as children's toys. We are informed that schoolboys played with them everywhere, sometimes by balancing four of the astragali on the knuckles of a hand, tossing them by a flip of the hand into the air, and then endeavoring to recapture them as they fell. Also, from Greek vase-painting, the astragali were sometimes tossed into a ring drawn on the ground, much as children of today play with marbles. Whether man adopted the toys of his children or the children copied the man is impossible to say, but by the First Dynasty in Egypt (about 3500 B.C.), astragali were used in a variety of games, in some of which "men" were moved about on a board according to the fall of a tossed astragalus bone. There is an Egyptian tomb-painting showing a nobleman in after-life with a playing board set out before him and an astragalus delicately balanced on his finger tip prior to being tossed. Children of today in France and Italy still play games with astragalus bones, and metallic versions of the bones can be purchased in village shops.

This is not the place to enter into the shadowy history of game-playing, nor into the cloudy origin of *gaming*. Did gaming develop from game-playing, or did it arise from wagering and the drawing of lots, or from religious divination and the consultation of oracles? In any case, by approximately 1200 B.C. the cubical marked die had evolved as a more suitable randomizing agent in games than the astragalus. This idealization occurred simultaneously in different parts of the world, and it is quite likely that the first primitive dice were made by rubbing flat the two opposite rounded faces of astragalus bones. The faces of a die were variously marked by drilling into them numbers of small shallow depressions with some sort of a circular engraving tool.

It was natural that gaming, as a game using only dice with no accompanying playing board and pieces, should arise and that players should concern themselves with the chances, or probabilities, of obtaining sums with the throw of two or more dice. Thus, although Greek philosophers of antiquity discussed necessity and contingency at some length, it is perhaps correct to say that the beginnings of a study of probability arose in attempts to evaluate the chances in certain gambling games, particularly the game of dice.

It is hard for historians of the calculus of probability to account for the extremely tardy conceptual growth of the subject. Of course, a realization of the equally likely possibilities in dice-throwing would of necessity be delayed until "honest" dice were made. So long as astragalus bones, or simple handy pieces of wood, ivory, or stone, smoothed off and appropriately marked, were used for either play or divination, the regularity of fall of the different faces would be quite obscured. Also, long series of trials are needed to calculate empirical probabilities, and there would have been few persons capable of keeping a tally of throws and of making the required enumerations. There seemed little alternative to the feeling that the fall of dice or astragali was completely controlled by the whimsies of the gods.

We know that a passion for gaming possessed the Roman Emperors and the surrounding leisured rich. It is said, for example, that Claudius (10 B.C.–A.D. 54) was greatly devoted to dicing and had even published a book, which unfortunately has not survived, entitled *How to Win at Dice*. But a real start in the calculation of random events did not take place until the Renaissance, when the ability to write and calculate with numbers had become widespread and simple algebra had developed.

It seems proper to say that there was no truly mathematical treatment of probability until the latter part of the fifteenth century and the early part of the sixteenth century, when some of the Italian mathematicians attempted to evaluate the chances in certain gambling games, like that of dice. Girolamo Cardano (1501–1576), as was noted in LECTURE 16, wrote a brief gambler's manual in which some of the simpler aspects of mathematical probability are involved. But it is generally agreed that the one problem to which can be credited the origin of the science of probability is the so-called *problem of the points*. This problem requires the determination of the division of the stakes of an interrupted game of chance between two supposedly equally skilled players, knowing the scores of the players at the time of interruption and the number of points needed to win the game. Fra Luca Pacioli (1445–1509), in his popular *Sūma** of 1494, was one of the first writers to introduce the problem of the points into a work on mathematics. The problem was subsequently discussed by Cardano and Tartaglia (ca. 1499–1557). All these men arrived at incorrect answers. A real advance was not made until the problem was proposed, in 1654, to Blaise Pascal, by the Chevalier de Méré, an able and experienced gambler whose theoretical reasoning on the problem did not agree with his observations. Pascal became interested in the problem and communicated it to Pierre de Fermat. There ensued a remarkable correspondence between these two French mathematicians,† in which the problem was correctly but differently solved by each man. It was in this correspondence of 1654 that Pascal and Fermat jointly laid the foundations of the theory of mathematical probability—a GREAT MOMENT IN MATHEMATICS had arrived.

---

*More completely, *Summa de arithmetica, geometria, proportioni e proportionalità*.
†This correspondence appears in D. E. Smith, *A Source Book in Mathematics*.

Blaise Pascal was born in 1623 in the French province of Auvergne and very early showed exceptional ability in mathematics. When only 12 he discovered, entirely on his own, many of the theorems of elementary plane geometry. At 14 he took part in the informal weekly sessions of a group of mathematicians from which the French Academy eventually arose in 1666. At 16 he discovered, among other things, his singularly rich "mystic hexagram" theorem* of projective geometry. A few years later he invented and constructed the first adding machine and began to apply his unusual talents to physics and mechanics. In 1648 he wrote a comprehensive, but now lost, treatise on projective geometry.

This astonishing and precocious activity came to a sudden halt in 1650, when, suffering from fragile health, Pascal decided to abandon his work in mathematics and science and to devote himself to religious meditation. Three years later, however, he transitorily returned to mathematics, at which time he wrote his *Traité du triangle arithmétique*, which, as we shall shortly see, played an important part in the matter that concerns us in the present lecture. He conducted a number of experiments on fluid pressure, which led to the invention of the hydraulic press, and, in 1654, carried on the historic correspondence with Fermat that laid the foundations of the mathematical theory of probability.

Then, late in 1654, Pascal received what he felt to be a strong intimation that his renewed activities in mathematics and science were displeasing to God. The divine hint occurred when his runaway horses dashed to their deaths over the high parapet of the bridge at Neuilly, and his own life was miraculously preserved only by the last minute breaking of the traces. Morally fortified by a reference to the accident written on a small piece of parchment henceforth carried next to his heart, he dutifully returned to his religious contemplations.

It was only once again, in 1658, that Pascal reverted to mathematics. While suffering from excruciating toothache, some geometrical ideas occurred to him, and his teeth forthwith ceased to ache. Interpreting this as a sign of divine will, he assiduously applied himself for eight days expanding his ideas, producing a fairly complete account of the geometry of the cycloid curve.

---

*The three points of intersection of the three pairs of opposite sides of any hexagon inscribed in any conic are collinear.

Pascal's famous *Provincial Letters* and his *Pensées,* both dealing with religious matters and read today as models of early French literature, were composed toward the close of his brief life. He died in Paris, after a lingering and complicated illness, in 1662 at the pathetically young age of 39.

Pascal has been called the greatest "might-have-been" in the history of mathematics. Possessing such remarkable talents and such keen geometrical intuition, he should have produced a great deal more. Unfortunately, much of his life was spent suffering the racking physical pains of acute neuralgia and the distressing mental torments of religious neuroticism.

In contrast to the short, disturbed, tortured, and only spasmodically productive life of Blaise Pascal, Pierre de Fermat's life was moderately long, peaceful, enjoyable, and almost continuously productive. Fermat was born at Beaumont de Lomagne, near Toulouse, in 1601(?), as the son of a well-to-do leather merchant. He received his early education at home, as did Pascal.

In 1631, Fermat was installed at Toulouse as commissioner of requests, and in 1648 was promoted to the post of King's councilor to the local parliament at Toulouse. In this latter capacity he spent the rest of his life discharging his duties with modesty and punctiliousness. While thus serving as a humble and retiring lawyer, he devoted the bulk of his leisure time to the study and creation of mathematics. Although he published very little during his lifetime, he was in scientific correspondence with many of the leading mathematicians of his day, and in this way considerably influenced his contemporaries.

Fermat enriched so many branches of mathematics with so many important contributions that he has been called the greatest French mathematician of the seventeenth century. We have seen in our preceding lecture that he was an independent inventor of analytic geometry; in the present lecture we shall see how he helped lay the foundations of the mathematical theory of probability, and in our next lecture we shall see that he contributed noteworthily to the early development of the calculus. But of all his varied contributions to mathematics, by far the most outstanding is his founding of the modern theory of numbers, a field in which he possessed extraordinary intuition and an awesomely impressive ability, putting him among the top number theorists of all time.

Fermat died in Castres (or perhaps Toulouse), quite suddenly, in

1665. His tombstone, originally in the Church of the Augustines in Toulouse, was later moved to the local museum.

[Let us now turn to the problem of the points, the solutions of which by Pascal and Fermat, in their correspondence of 1654, commenced a sound mathematical study of probability.] An illustrative case discussed by the two French mathematicians was that in which one seeks the division of the stakes in a game of chance between two equally skilled players A and B where player A needs 2 more points to win and player B needs 3 more points to win. We first consider Fermat's solution to the problem, since it is the simpler and more direct of the two; and then Pascal's solution, which perhaps is more refined and more capable of generalization.

Inasmuch as it is clear, in the illustrative example, that four more trials will decide the game, Fermat let a indicate a trial where A wins and b a trial where B wins, and considered the 16 possible permutations of the two letters a and b taken 4 at a time:

$$
\begin{array}{llll}
aaaa & aaab & abba & bbab \\
baaa & bbaa & abab & babb \\
abaa & baba & aabb & abbb \\
aaba & baab & bbba & bbbb
\end{array}
$$

The cases where a appears 2 or more times are favorable to A; there are 11 of them. The cases where b appears 3 or more times are favorable to B; there are 5 of them. Therefore the stakes should be divided in the ratio 11:5. For the general case, where A needs m points to win and B needs n, one writes down the 2^{m+n-1} possible permutations of the two letters a and b taken $m + n - 1$ at a time. One then finds the number α of cases where a appears m or more times and the number β of cases where b appears n or more times. The stakes are then to be divided in the ratio $\alpha : \beta$.

Pascal solved the problem of the points by utilizing his "arithmetic triangle," an array of numbers discussed by him in his *Traité du triangle arithmétique*, which, though not published until 1665, was written in 1653. He constructed his "arithmetic triangle" as indicated in Figure 1. Any element (in the second or a following row) is obtained as the sum of all those elements of the preceding row lying just above or to the left of the desired element. Thus, in the fourth row,

$$35 = 15 + 10 + 6 + 3 + 1.$$

The triangle, which may be of any order, is obtained by drawing a diagonal as shown in the figure. The student of college algebra will recognize that the numbers along such a diagonal are the successive coefficients in a binomial expansion. For example, the numbers along the fifth diagonal, namely 1, 4, 6, 4, 1, are the successive coefficients in the expansion of $(a + b)^4$. The finding of binomial coefficients was one of the uses to which Pascal put his triangle. He also used it for finding the number of combinations of n things taken r at a time, which he correctly stated to be

$$C(n, r) = n!/r!(n - r)!,$$

where $n!$ is our present-day notation* for the product

$$n(n - 1)(n - 2) \cdots (3)(2)(1).$$

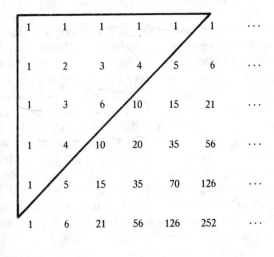

Fig. 1

---

*The symbol $n!$, called *factorial n*, was introduced in 1808 by Christian Kramp (1760–1826) of Strasbourg, who chose this symbol so as to circumvent printing difficulties incurred by a previously used symbol. For convenience one defines $0! = 1$.

One can easily show that the elements along the fifth diagonal are, respectively,

$$C(4, 4) = 1, \qquad C(4, 3) = 4, \qquad C(4, 2) = 6,$$
$$C(4, 1) = 4, \qquad C(4, 0) = 1.$$

Since $C(4, 4)$ is the number of ways to obtain 4 a's, $C(4, 3)$ is the number of ways to obtain 3 a's, etc., it follows that the solution of the illustrative problem of the points is given by

$$[C(4, 4) + C(4, 3) + C(4, 2)] : [C(4, 1) + C(4, 0)]$$
$$= (1 + 4 + 6) : (4 + 1) = 11 : 5.$$

In the general case, where A needs m points to win and B needs n, one chooses the $(m + n)$th diagonal of Pascal's arithmetic array. One then finds the sum α of the first n elements of this diagonal and the sum β of the last m elements. The stakes are to be divided in the ratio $\alpha : \beta$.

There are many relations involving the numbers of the arithmetic triangle, several of which were developed by Pascal. Pascal was not the originator of the arithmetic triangle, for such an array had been anticipated several centuries earlier by Chinese and Persian writers and had been considered by a number of Pascal's European predecessors. It is because of Pascal's development of many of the triangle's properties and because of the applications which he made of these properties that the array has become known as *Pascal's triangle*. In Pascal's treatise on the triangle appears one of the earliest acceptable statements of the method of mathematical induction.

Pascal and Fermat, in their historic correspondence of 1654, reflected upon other problems related to the problem of the points, such as the division of stakes where there are more than two players, or where there are two unevenly skilled players. With this work by Pascal and Fermat, marking a GREAT MOMENT IN MATHEMATICS, the mathematical theory of probability was well launched. In 1657, the great Dutch genius Christiaan Huygens (1629–1695) wrote the first formal treatise on probability, basing his work on the Pascal-Fermat correspondence. This was the best account of the subject until the posthumous appearance, in 1713, of the *Ars conjectandi* of Jacob Bernoulli (1654–1705), which contained a reprint of the earlier treatise by Huygens. After these pioneering efforts, we find the subject carried forward by such men as Abraham De Moivre (1667–1754), Daniel Bernoulli (1700–1782), Leonhard Euler (1707–1783), Joseph Louis Lagrange (1736–1813), Pierre-Simon Laplace (1749–1827), and a host of other contributors.

It is fascinating, and at the same time somewhat astonishing, to contemplate that mathematicians have been able to develop a science, namely the mathematical theory of probability, that establishes rational laws that can be applied to situations of pure chance. This science is far from being impractical, as is attested by experiments performed in great laboratories, by the existence of highly respected insurance companies, and by the logistics of big businesses and of war. About this science of probability, the eminent French mathematician Pierre-Simon Laplace remarked that, though it started with the consideration of certain lowly games of chance, it rose to become one of the most important areas of human knowledge. The great British mathematical-physicist, James Clerk Maxwell (1831–1879), claimed that it is the "mathematics for practical men." And the English logician and economist William Stanley Jevons (1835–1882) said it is "the very guide of life and hardly can we take a step or make a decision without correctly or incorrectly making an estimate of probability."

Exercises

21.1. Find the division of the stakes in a game of chance between two equally skilled players A and B where

(a) A needs 1 more point to win and B needs 4 more points to win, using Fermat's enumeration method.

(b) A needs 3 more points to win and B needs 4 more points to win, using Pascal's triangle method.

21.2. Show that
(a) $C(n, r) = n!/r!(n - r)!$
(b) $C(n, n - r) = C(n, r)$
(c) $C(n, r) = C(n - 1, r) + C(n - 1, r - 1)$.

21.3. (a) Show that the coefficient of $a^{n-r}b^r$ in the expansion of $(a + b)^n$ is $C(n, r)$.

(b) Show that $C(n, 0) + C(n, 1) + C(n, 2) + \cdots + C(n,n) = 2^n$.

21.4. Establish the following relations, all of which were developed by Pascal, involving the numbers of the arithmetic triangle.

(a) Any element (not in the first row or the first column) of the arithmetic triangle is equal to the sum of the element just above it and the element just to the left of it.

(b) Any given element of the arithmetic triangle, decreased by 1, is equal to the sum of all the elements above the row and to the left of the column containing the given element.

(c) The mth element in the nth row is $C(m + n - 2, n - 1)$.

(d) The element in the mth row and nth column is equal to the element in the nth row and mth column.

(e) The sum of the elements along any diagonal is twice the sum of the elements along the preceding diagonal.

(f) The sum of the elements along the nth diagonal is 2^{n-1}.

(g) Show that $C(n, r)$ appears at the intersection of the $(n + 1)$st diagonal and the $(r + 1)$st column of the arithmetic triangle.

Further Reading

BELL, E. T., *Men of Mathematics*. New York: Simon and Schuster, 1937.

DAVID, F. N., *Games, Gods and Gambling*. New York: Hafner, 1962.

SMITH, D. E., *A Source Book in Mathematics*. New York: Dover, 1958.

TODHUNTER, I., *A History of the Mathematical Theory of Probability from the Time of Pascal to that of Laplace*. New York: Chelsea, 1949.

MOVING PICTURES VERSUS STILL PICTURES

The prime stimulus to the invention of new mathematical procedures is the presence of problems whose solutions have evaded known methods of mathematical attack. Indeed, the continual appearance of unsolved problems constitutes the life blood that maintains the health and growth of mathematics. In our previous lecture we saw an example of this—it was an elusive problem, the so-called *problem of the points,* that led to the creation of the field of mathematical probability.

In earlier lectures we have seen that the problem of finding certain areas, volumes, and arc lengths gave rise to summation processes that led to the creation of the integral calculus. In the present lecture we shall see that the problem of drawing tangents to curves and the problem of finding maximum and minimum values of functions led to the creation of the differential calculus. Each of these creations certainly constitutes a GREAT MOMENT IN MATHEMATICS.

It is interesting that, whereas the origins of the integral calculus go back to classical Greek times, it is not until the seventeenth century that we find significant contributions to the differential calculus. Not that there was no prior attempt at drawing tangents to curves and no prior employment of maximum and minimum considerations. For example, the Greeks of antiquity were able to draw tangents to circles and to the conic sections. Apollonius, in his *Conic Sections,* treated normals to a conic as the maximum and minimum line segments drawn from a point to the curve, and other maximum and minimum considerations can be found in the works of the ancient Greeks. Again, many centuries later, something of a more general approach to drawing tangents to curves was given by Gilles Persone de Roberval (1602–1675). He endeavored to consider a curve as

11

generated by a point whose motion is compounded from two known motions. Then the resultant of the velocity vectors of the two known motions gives the tangent line to the curve. For example, in the case of a parabola, we may consider the two motions as away from the focus and away from the directrix. Since the distances of the moving point from the focus and the directrix are always equal to each other, the velocity vectors of the two motions must also be of equal magnitude. It follows that the tangent at a point of the parabola bisects the angle between the focal radius to the point and the perpendicular through the point to the directrix (see Figure 2). This idea of tangents was also held by Evangelista Torricelli (1608–1647), and an argument of priority of invention ensued between Roberval and Torricelli. Attractive as the method is, however, it seems quite limited in application.

Another method of constructing tangents to certain curves was given by René Descartes in the second part of his *La géométrie* of 1637. Though he applied his method to a number of different

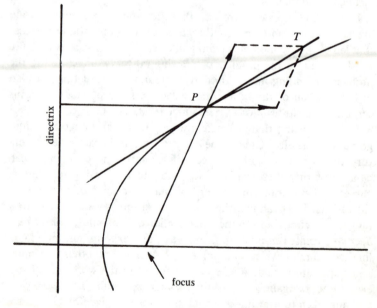

FIG. 2

curves, including one of the quartic ovals named after him,* the method is restricted to algebraic curves and, even at that, too often leads to forbidding algebra.

None of the methods above has general application, nor does any of them contain the procedure of differentiation. The first really marked anticipation of differentiation stems from ideas set forth by Fermat in 1629, though not much publicized until some eight or nine years later. Kepler had observed that the increment of a function becomes vanishingly small in the neighborhood of an ordinary maximum or minimum value. Fermat translated this fact into a process for determining such a maximum or minimum. In brief his method is this. If $f(x)$ has an ordinary maximum or minimum at x, and if e is very small, then the value of $f(x + e)$ is almost equal to that of $f(x)$. Therefore, we tentatively set $f(x + e) = f(x)$ and then make the equality correct by letting e assume the value zero. The roots of the resulting equation then give those values of x for which $f(x)$ is a maximum or a minimum.

Let us illustrate and clarify this procedure by considering Fermat's first example—to divide a quantity into two parts such that their product is a maximum. Fermat used Viète's notation, where constants are designated by upper-case consonants and variables by upper-case vowels. Employing this notation, let B be the given quantity and denote the desired parts by A and $B - A$. Forming the product

$$(A + E)[B - (A + E)]$$

and equating it to $A(B - A)$ we have

$$A(B - A) = (A + E)(B - A - E)$$

or

$$BE - 2AE - E^2 = 0.$$

After dividing by E, one obtains

$$B - 2A - E = 0.$$

---

*A *Cartesian oval* is the locus of a point whose distances, r_1 and r_2, from two fixed points satisfy the relation $r_1 + mr_2 = a$, where m and a are constants. The central conics will be recognized as special cases.

Now setting $E = 0$ we obtain $2A = B$, and thus find that the required division demands that the two parts each be half of B.

Although the logic of Fermat's exposition leaves much to be desired, it is seen that his method is equivalent to setting

$$\lim_{h \to 0} \frac{f(x + h) - f(x)}{h} = 0,$$

that is, to setting the derivative of $f(x)$ equal to zero. This is the customary method for finding ordinary maxima and minima of a function $f(x)$, and is sometimes referred to in our elementary textbooks as *Fermat's method*. Fermat, however, did not realize that the vanishing of the derivative of $f(x)$ is only a necessary, but not a sufficient, condition for an ordinary maximum or minimum. Also, Fermat's method does not distinguish between a maximum and a minimum value.

Fermat also devised a general procedure for finding the tangent at a point of a curve whose Cartesian equation is given. His idea is to find the *subtangent* for the point, that is, the segment on the x-axis between the foot of the ordinate drawn to the point of contact and the intersection of the tangent line with the x-axis. The method employs the idea of a tangent as the limiting position of a secant when two of the points of intersection of the secant with the curve tend to coincide. Using modern notation the method is as follows. Let the equation of the curve (see Figure 3) be $f(x,y) = 0$, and let us seek the subtangent t of the curve for the point (x, y) of the curve. By similar triangles we easily find the coordinates of a near point on the tangent to be

$$\left[x + e, y \left(1 + \frac{e}{t} \right) \right].$$

This point is tentatively treated as if it were also on the curve, giving us

$$f \left[x + e, y \left(1 + \frac{e}{t} \right) \right] = 0.$$

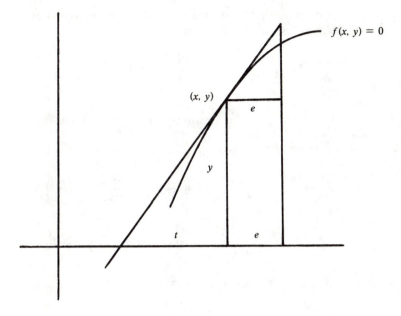

$f(x, y) = 0$

(x, y)

e

y

t

e

FIG. 3

The equality is then made correct by letting e assume the value zero. We then solve the resulting equation for the subtangent t in terms of the coordinates x and y of the point of contact. This, of course, is equivalent to setting

$$t = -y \frac{\frac{\partial f}{\partial y}}{\frac{\partial f}{\partial x}},$$

a general formula that appeared later in 1652, naturally without the modern notation, in the work of René François Walter de Sluze (1622–1685), a canon in the Church who wrote numerous tracts in mathematics. Fermat, using his method, found tangents to the ellipse, cycloid, cissoid, conchoid, quadratrix, and folium of Descartes. Let us illustrate the method by finding the subtangent at a general point on the folium of Descartes:

$$x^3 + y^3 = nxy.$$

Here we have

$$(x + e)^3 + y^3 \left(1 + \frac{e}{t}\right)^3 - ny(x + e)\left(1 + \frac{e}{t}\right) = 0,$$

or

$$e\left(3x^2 + \frac{3y^3}{t} - \frac{nxy}{t} - ny\right) + e^2\left(3x + \frac{3y^3}{t^2} - \frac{ny}{t}\right)$$

$$+ e^3\left(1 + \frac{y^3}{t^3}\right) = 0.$$

Now, dividing by e and then setting $e = 0$, we find

$$t = -\frac{3y^3 - nxy}{3x^2 - ny}.$$

Another man who played a part in anticipating differentiation was Isaac Barrow. Barrow was born in London in 1630 and completed his education at Cambridge University. He was a man of high academic caliber, achieving recognition in mathematics, physics, astronomy, and theology. He was also renowned as one of the best Greek scholars of his day. He was the first to occupy the Lucasian chair at Cambridge, from which he magnanimously resigned in 1669 in favor of his great pupil, Isaac Newton, whose remarkable abilities he was one of the first to recognize and acknowledge. He died in Cambridge in 1677.

Barrow's most important mathematical work is his *Lectiones opticae et geometricae*, which appeared in the year he resigned his chair at Cambridge. The preface to the treatise admits indebtedness to Newton for some of the material of the book, probably the parts dealing with optics. It is in this book that we find a very near approach to the modern process of differentiation, utilizing essentially the so-called *differential triangle* which we find in our present-day calculus textbooks. Let it be required to find the tangent at a point P on the given curve represented in Figure 4. Let Q be a neighboring point on the curve. Then triangles PTM and PQR are very nearly similar to one another, and, Barrow argued, as the little triangle becomes indefinitely small, we have

$$RP/QR = MP/TM.$$

Let us set $QR = e$ and $RP = a$.* Then if the coordinates of P are x and y, those of Q are $x - e$ and $y - a$. Substituting these values in

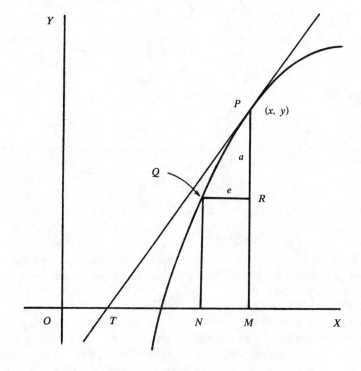

Fig. 4

---

*It is to be noted that a and e are the Δy and Δx of present-day treatments, whence the ratio a/e becomes dy/dx when $e \to 0$.

the equation of the curve and neglecting squares and higher powers of both e and a, we find the ratio a/e. We then have

$$OT = OM - TM = OM - MP(QR/RP) = x - y(e/a),$$

and the tangent line is determined. Barrow applied this method of constructing tangents to the curves:

(a) $x^2(x^2 + y^2) = r^2y^2$ (the *kappa curve*),
(b) $x^3 + y^3 = r^3$ (a special *Lamé curve*),
(c) $x^3 + y^3 = rxy$ (the *folium of Descartes,* but called *la galande* by Barrow),
(d) $y = (r - x)\tan(\pi x/2r)$ (the *quadratrix*),
(e) $y = r\tan(\pi x/2r)$ (a *tangent curve*).

As an illustration, let us apply the method to the curve (b). Here we have

$$(x - e)^3 + (y - a)^3 = r^3,$$

or

$$x^3 - 3x^2e + 3xe^2 - e^3 + y^3 - 3y^2a + 3ya^2 - a^3 = r^3.$$

Neglecting the square and higher powers of e and a, and using the fact that $x^3 + y^3 = r^3$, this reduces to

$$3x^2e + 3y^2a = 0,$$

from which we obtain

$$a/e = -x^2/y^2.$$

This ratio a/e is, of course, our modern dy/dx, and Barrow's dubious procedure can easily be made rigorous by use of the theory of limits.

With the work of Fermat, Barrow, and some of their contemporaries, a process of differentiation had been evolved and applied to the resolution of a number of maxima and minima problems and to the construction of tangents to many curves. What more in the development of the differential calculus remained to be done? There still remained the creation of a general symbolism with a systematic set of formal analytical rules for the calculation of derivatives, and also a consistent and rigorous redevelopment of the foundations of

the subject. It is precisely the first of these, the creation of a suitable
and workable *calculus,* that was furnished by Newton and Leibniz,
working independently of one another. The redevelopment of the
fundamental concepts on an acceptably rigorous basis had to outwait the period of energetic application of the subject, and was the
work of the great French analyst Augustin-Louis Cauchy (1789–
1857) and his nineteenth-century successors. This story, which is
another GREAT MOMENT IN MATHEMATICS, will be told in a later lecture.

The first to publish a general and workable differential calculus
was the great German mathematician and philosopher Gottfried
Wilhelm Leibniz (1646–1716). In the journal *Acta eruditorum** of
1684, in an article entitled "A new method for maxima and minima
as well as tangents, which is not restricted by fractional or irrational
quantities, and a remarkable type of calculus for this," Leibniz
published a concise exposition of his differential calculus, the formulation of which he says dated from 1676. In spite of several obscure points and some careless errors, the paper proved to be a land-
mark in the further advancement of mathematics. The notation of
the differential calculus and many of the general rules for calculating derivatives that are in use today were given by Leibniz in this
paper. Leibniz wrote as follows:

> Let an axis *AX* [see Figure 5, which is Leibniz's figure simplified and
> slightly augmented] and several curves *VV, WW, YY, ZZ* be given, of
> which the ordinates *VX, WX, YX, ZX,* perpendicular to the axis, are
> called *v, w, y, z* respectively. The segment *AX* cut off from the axis is
> called *x.* Let the tangents be *VB, WC, YD, ZE,* respectively intersect-
> ing the axis at *B, C, D, E.* Now some arbitrarily selected segment is
> called *dx,* and the line segment which is to *dx* as *v* (or *w,* or *y,* or *z*) is
> to *XB* (or *XC,* or *XD,* or *XE*) is called *dv* (or *dw,* or *dy,* or *dz*), for the
> difference of these *v* (or *w,* or *y,* or *z*).

Leibniz then goes on to derive a number of familiar differentiation
rules, such as:

(1) If *a* is a constant, then $da = 0$.

---

*This journal was founded in 1682 by Leibniz and Otto Mencke, with Leibniz serv-
ing as editor-in-chief.

(2) $d(ax) = a\,dx.$

(3) $d(w - y + z) = dw - dy + dz.$

(4) $d(x^n) = nx^{n-1}dx$ (n a natural number).

(5) $d(1/x^n) = -\dfrac{n\,dx}{x^{n+1}}.$

(6) $d(^b\sqrt{x^a}) = \dfrac{a}{b}\ ^b\sqrt{x^{a-b}}\,dx.$

(7) $d(vy) = v\,dy + y\,dv.$

(8) $d(v/y) = \dfrac{y\,dv - v\,dy}{y^2}.$

This is Leibniz's differential calculus, which makes differentiation an almost mechanical operation, whereas previously one had to go through the limiting procedure in each individual case. Further-

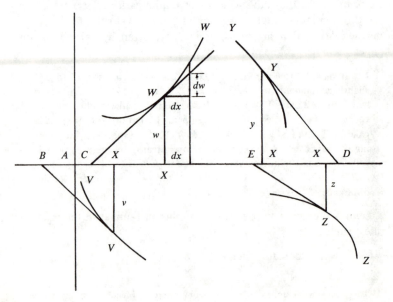

FIG. 5

more, Leibniz had introduced a fortunate, well-devised, and highly satisfying symbolism. The *differential calculus* had been invented.*

Leibniz had a happy knack for choosing convenient symbolism. Not only did he give us our present-day supple notation of the differential calculus, but in 1675 he introduced the modern integral sign, as a long letter *S* derived from the first letter of the Latin word *summa* (sum), to indicate the sum of Cavalieri's indivisibles. Though the invention of our symbolism of the differential calculus is to be credited solely to Leibniz, the invention of the calculus itself must be shared with the preëminent British mathematician and physicist Isaac Newton (1643–1727). As a matter of fact, Newton devised his *fluxional calculus,* as he called it, earlier than Leibniz had devised his *differential calculus,* but he did not publish his work until 1687. This delay of publication by Newton led to the greatest quarrel on priority of discovery in the history of mathematics.

The facts of the case are these. Newton developed his fluxional calculus as early as 1665, with the initial intention that it be applied to problems in physics, and only a few close colleagues knew of his creation. Many years later, in a letter to Leibniz sent via Henry Oldenburg, the secretary of the British Royal Society, Newton briefly and somewhat obscurely described his method, whereupon Leibniz, who by that time had already developed his own method, in a reply described his calculus to Newton. There the interchange of correspondence ceased. In the following years, Leibniz's differential calculus spread, by word of mouth, among the leading mathematicians of continental Europe, who applied it to many different problems with outstanding success. But not until 1684 did Leibniz actually put his invention into print, in the paper cited above and with no mention of Newton's corresponding achievement. Consequently Newton mentioned, in a scholium in his great scientific treatise *Philosophiae naturalis principia mathematica* of 1687, the exchange of letters that had earlier taken place. Thereupon contemporaries and successors of Newton and Leibniz started a priority quarrel, with accusations and counteraccusations of plagiarism, that often became highly undignified and degenerated into a political dispute

---

*We are not here concerned with any epistemological distinction between *discovery* and *invention.*

between England and Germany. So great was the English national pride that, a century or so after the quarrel, the English mathematicians steadfastly stuck to Newtonian terminology and symbolism, much to the detriment of British mathematics. Historical research has concluded that Newton and Leibniz, traveling different routes, each on his own arrived at essentially the same goal, and therefore the two men are to be regarded as independent inventors of the differential calculus.

Newton's approach to the calculus was a physical one. He considered a curve as generated by the continuous motion of a point. It follows that the abscissa and the ordinate of the generating point are, in general, changing quantities. He called a changing quantity a *fluent* (a flowing quantity), and its rate of change he called the *fluxion* of the fluent. If a fluent, such as the ordinate of the point generating a curve, is represented by y, then the fluxion of the fluent is represented by $\dot{y}$. In modern notation we see that this is equivalent to dy/dt, where t represents time. The fluxion of $\dot{y}$ is denoted by $\ddot{y}$, and so on, for higher-ordered fluxions. On the other hand, the fluent of y is denoted by the symbol y with a small square drawn about it, or sometimes by y. Newton also introduced another concept, which he called the *moment* of the fluent; it is the infinitely small amount by which a fluent increases in an infinitely small time interval o. Thus the moment of the fluent x is given by the product $\dot{x}o$. Newton remarked that we may, in any problem, neglect all terms that are multiplied by the second or higher power of o, and thus obtain an equation between the coordinates x and y of the generating point of a curve and their fluxions $\dot{x}$ and $\dot{y}$.

As an illustration of Newton's method, let us consider an example given by Newton in his work *Method of Fluxions and Infinite Series,* written in 1671 but not published until 1736, nine years after he had died. Here Newton considers the cubic curve

$$x^3 - ax^2 + axy - y^3 = 0.$$

Replacing x by $x + \dot{x}o$ and y by $y + \dot{y}o$, we get

$$\begin{aligned}
&x^3 + 3x^2(\dot{x}o) + 3x(\dot{x}o)^2 + (\dot{x}o)^3 \\
&- ax^2 - 2ax(\dot{x}o) - a(\dot{x}o)^2 \\
&+ axy + ay(\dot{x}o) + a(\dot{x}o)(\dot{y}o) + ax(\dot{y}o) \\
&- y^3 - 3y^2(\dot{y}o) - 3y(\dot{y}o)^2 - (\dot{y}o)^3 = 0.
\end{aligned}$$

Now, using the fact that $x^3 - ax^2 + axy - y^3 = 0$, dividing the remaining terms by o, and then rejecting all terms still containing o as a factor, we find

$$3x^2\dot{x} - 2ax\dot{x} + ay\dot{x} + ax\dot{y} - 3y^2\dot{y} = 0.$$

If, in the last equation, we should divide by $\dot{x}$ and then solve for $\dot{y}/\dot{x}$, we would find

$$\dot{y}/\dot{x} = (3x^2 - 2ax + ay)/(3y^2 - ax).$$

Of course, in our modern notation,

$$\dot{y}/\dot{x} = (dy/dt)/(dx/dt) = dy/dx.$$

The rejection of terms containing the second and higher powers of o was much criticized by some of Newton's contemporaries. Newton later justified the process by introducing the notion of *ultimate ratios,* which is a primitive conception of the limit idea.

Newton considered two types of problems. In the first type we are given a relation connecting some fluents, and we are asked to find a relation connecting these fluents and their fluxions. This is what we did above, and is, of course, equivalent to differentiation. In the second type we are given a relation connecting some fluents and their fluxions, and we are asked to find a relation connecting the fluents alone. This is the inverse problem and is equivalent to solving a differential equation. Newton made numerous and remarkable applications of his method of fluxions. He determined maxima and minima, tangents to curves, curvature of curves, points of inflection, and convexity and concavity of curves, and he applied his theory to numerous quadratures and to rectification of curves. In the integration of some differential equations he showed extraordinary ability.

The creation of the differential calculus marks a watershed, or turning point, in the history of mathematics. The new mathematics inspired by this great invention differs markedly from the old mathematics that had been largely inherited from the ancient Greeks. The older mathematics appears static while the newer appears dynamic, so that the older mathematics compares to the still-picture stage of photography while the newer mathematics compares to the moving-picture stage. Again, the older mathematics is to the newer as anatomy is to physiology, wherein the former studies the dead body and the latter studies the living body. Once more, the older mathematics concerned itself with the fixed and the finite while the newer mathematics embraces the changing and the infinite.

Needless to say, the creation of the differential calculus was a truly GREAT MOMENT IN MATHEMATICS, and, to be fair to all involved, we perhaps should assign it to the period running from the initiatory efforts of Fermat in 1629 through the epoch-making work of Newton and Leibniz consummated over fifty years later. We shall have more to say about the calculus in our next lecture.

Exercises

22.1. Apply Roberval's method to the drawing of tangents to (a) an ellipse, (b) a hyperbola.

22.2. Following is Descartes' method of drawing tangents (see Figure 6). Let the equation of the given curve be $f(x, y) = 0$ and let (x_1, y_1) be the coordinates of the point P of the curve at which we wish to construct a tangent. Let Q, having coordinates $(x_2, 0)$, be a point on the x-axis. Then the equation of the circle with center Q and radius QP is

$$(x - x_2)^2 + y^2 = (x_1 - x_2)^2 + y_1^2.$$

Eliminating y between this equation and the equation $f(x, y) = 0$ yields an equation in x leading to the abscissas of the points where the circle cuts the given curve. Now determine x_2 so that this equation in x will have a *pair* of roots equal to x_1. This condition fixes Q as the intersection of the x-axis and the normal to the curve at P, since the circle is now tangent to the given curve at P. The required tangent is the perpendicular through P to PQ.

Construct, by Descartes' method, the tangent to the parabola $y^2 = 4x$ at the point $(1,2)$.

22.3. Show that the slope of the tangent to the curve $y = f(x)$ at the point having abscissa x_1 is given by $f'(x_1)$, where $f'(x)$ denotes the derivative of $f(x)$.

22.4. Find the slope of the tangent at the point $(3, 4)$ on the circle $x^2 + y^2 = 25$ by:
    (a) Fermat's method,

(b) Barrow's method,

(c) Newton's method of fluxions,

(d) the method taught in calculus classes today.

22.5. The following procedure is known as the *four-step rule*, or *ab initio process*, for finding the derivative of a given function $y = f(x)$.

    I. In $y = f(x)$, replace x by $x + \Delta x$, letting y become $y + \Delta y$.

    II. Subtract the original relation to obtain

$$\Delta y = f(x + \Delta x) - f(x).$$

III. Divide both sides by Δx, to obtain

$$\frac{\Delta y}{\Delta x} = \frac{f(x + \Delta x) - f(x)}{\Delta x}.$$

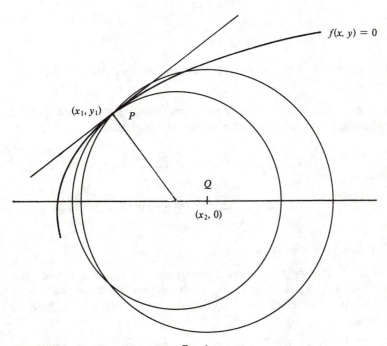

Fig. 6

IV. Take the limit of both sides as $\Delta x \rightarrow 0$ to obtain

$$\frac{dy}{dx} = \lim_{\Delta x \to 0} \frac{\Delta y}{\Delta x} = \lim_{\Delta x \to 0} \frac{f(x + \Delta x) - f(x)}{\Delta x}.$$

Using the four-step rule, obtain the following differentiation rules:

(a) If $y = a$, where a is a constant, then $dy/dx = 0$.

(b) If $y = ax$, then $dy/dx = a$.

(c) If $y = v - w + z$, where v, w, z are functions of x, then

$$dy/dx = dv/dx - dw/dx + dz/dx.$$

(d) If $y = x^n$, where n is a natural number, then $dy/dx = nx^{n-1}$.

(e) If $y = uv$, then $dy/dx = u(dv/dx) + v(du/dx)$.

(f) If $y = u/v$, then

$$\frac{dy}{dx} = \frac{v \dfrac{du}{dx} - u \dfrac{dv}{dx}}{v^2}.$$

(g) If $y = 1/x^n$, where n is a natural number, then $dy/dx = -n/x^{n+1}$.

22.6. If $y = uv$, where u and v are functions of x, show that the nth derivative, $y^{(n)}$, of y with respect to x is given by

$$y^{(n)} = uv^{(n)} + nu'v^{(n-1)} + \frac{n(n-1)}{2!} u''v^{(n-2)}$$

$$+ \frac{n(n-1)(n-2)}{3!} u'''v^{(n-3)} + \cdots + u^{(n)}v.$$

This is known as *Leibniz's rule*.

22.7. If $s = f(t)$, where s represents the distance a body has traveled along a straight-line path in time t, show that ds/dt measures the velocity and d^2y/dt^2 the acceleration of the body at time t.

22.8. Show that at an ordinary (turning-point) maximum or

minimum of a differentiable function $y = f(x)$ we must have $dy/dx = 0$. Show that, though this is a necessary condition, it is not a sufficient condition.

Further Reading

BOYER, C. B., *The History of the Calculus and Its Conceptual Development.* New York: Dover, 1959.

CHILD, J. M., *The Geometrical Lectures of Isaac Barrow.* Chicago: Open Court, 1916.

SMITH, D. E., *A Source Book in Mathematics.* New York: Dover, 1958.

STRUIK, D. J., *A Source Book in Mathematics, 1200–1800.* Cambridge, Mass.: Harvard University Press, 1969.

LIKE OPENING AND CLOSING A DOOR

It will be recalled that the integral calculus originated in the days of Greek antiquity in efforts to find certain areas, volumes, and arc lengths. The basic idea, in the case of areas, for example, is to consider an area as approximated by the areas of a great many very thin parallel rectangular strips. Using modern terminology, one then attempts to find the area as the limit approached by the sum of the areas of these strips, the number of the strips increasing indefinitely, and the widths of the strips approaching zero. With the ultimate introduction of analytic geometry in the seventeenth century, the matter finally assumed the following form.

Let $y = f(x)$ represent a continuous curve lying above the x-axis of a rectangular Cartesian coordinate system (see Figure 7). Consider the area A bounded by the x-axis, the given curve, and the ordinates to the curve erected at $x = a$ and $x = b$. Divide the area by ordinates into n vertical strips of respective widths Δx_1, Δx_2, ..., Δx_i, ..., Δx_n. Within each strip choose an ordinate, ordinate $f(x_1)$ within the first strip, ordinate $f(x_2)$ within the second strip, and so on. An approximation of the sought area A (consult Figure 7) is given by the sum of n thin rectangular areas, namely by

$$\sum_{i=1}^{n} f(x_i)\Delta x_i = f(x_1)\Delta x_1 + f(x_2)\Delta x_2 + \cdots$$

$$+ f(x_i)\Delta x_i + \cdots + f(x_n)\Delta x_n.$$

The area A itself is then given by

$$A = \lim_{\substack{n \to \infty \\ \text{each } \Delta x_i \to 0}} \sum_{i=1}^{n} f(x_i)\Delta x_i.$$

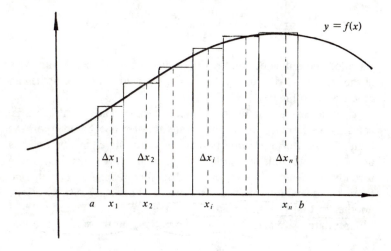

$$y = f(x)$$

Fig. 7

Carelessly stated, we have the area A given by the sum of an infinite number of infinitely thin rectangular elements of area. With the subsequent introduction of the integral sign by Leibniz, the area A came to be represented by the now familiar definite integral symbolism

$$A = \int_a^b f(x)\,dx.$$

That is, by definition,

$$\int_a^b f(x)\,dx = \lim_{\substack{n \to \infty \\ \text{each } \Delta x_i \to 0}} \sum_{i=1}^n f(x_i)\Delta x_i. \tag{1}$$

Here $f(x)\,dx$ originally represented the area of one of the infinitely thin elements of area, and the integral sign indicated that these elements are to be summed from the ordinate at $x = a$ to the ordinate at $x = b$.

In the seventeenth century the differential calculus was invented, and, as we learned in our last lecture, the derivative dF/dx of a function $y = F(x)$ became defined as

$$\frac{dF}{dx} = \lim_{\Delta x \to 0} \frac{\Delta y}{\Delta x} = \lim_{\Delta x \to 0} \frac{F(x + \Delta x) - F(x)}{\Delta x},$$

provided the indicated limit exists. This limit is, of course, the slope of the tangent to the curve $y = F(x)$ at the point on the curve having abscissa x (see Figure 8).

It would seem that in the integral calculus and the differential calculus we have two quite disparate studies, one based upon the limit of a sum of a growing number of vanishing elements, and the other upon the limit of a difference quotient. In the last half of the seventeenth century, a truly capital discovery was made, namely that these two seemingly disparate studies are intimately related to one another in that (for sufficiently restricted functions) integration and differentiation are actually inverse operations, in the same sense that

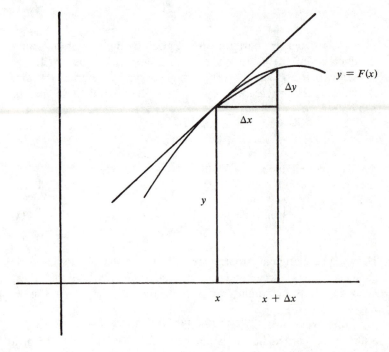

FIG. 8

addition and subtraction, or multiplication and division, are inverse operations. This astonishing fact is known as the *fundamental theorem of the calculus* and its discovery marks a GREAT MOMENT IN MATHEMATICS.

Let us briefly sketch a proof of the fundamental theorem of the calculus. Consider any definite integral

$$\int_a^b f(x)\,dx,$$

where $f(x)$ is continuous and nonnegative for all values of x in the number interval $a \leq x \leq b$. As seen above, this integral can be interpreted as the area bounded by the curve $y = f(x)$, the x-axis, and the ordinates at $x = a$ and $x = b$.

Now the area bounded by the curve, the x-axis, the ordinate at $x = a$, and a variable ordinate at an arbitrary point x of the interval $a \leq x \leq b$ (the obliquely shaded area in Figure 9) clearly depends upon x, is thus a function of x, and hence may be denoted by $A(x)$. We shall find

$$\frac{dA(x)}{dx}.$$

To this end let x be increased by an amount Δx, and denote by ΔA the corresponding increase in the area $A(x)$. It is evident from the figure that the area ΔA (the horizontally shaded area in Figure 9) lies between the areas of two vertical rectangles, each having Δx as base; one rectangle has the minimum y in Δx (call it min y) as altitude and the other has the maximum y in Δx (call it max y) as altitude. Therefore ΔA lies between (min $y)\Delta x$ and (max $y)\Delta x$. It is intuitively evident that there is a value $\bar{y}$ intermediate to min y and max y such that

$$\Delta A = \bar{y}\Delta x.$$

Hence $\Delta A/\Delta x = \bar{y}$. Now, as $\Delta x \to 0$, $\bar{y} \to y$, whence

$$\lim_{\Delta x \to 0} \frac{\Delta A}{\Delta x} = y.$$

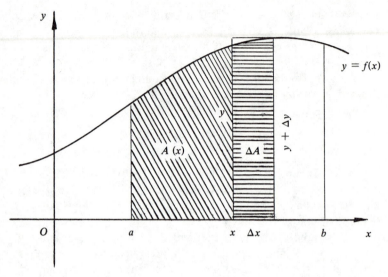

FIG. 9

That is, $dA/dx = y = f(x)$. It follows that $A(x)$ is a function such that its derivative is $f(x)$.

Of course, there is no unique function having a given derivative. If $F(x)$ has $f(x)$ for its derivative, then it can be shown that any other function having $f(x)$ for its derivative has the form $F(x) + C$, where C is some constant. Letting $F(x)$ be any particular function whose derivative is $f(x)$, it then follows that

$$A(x) = F(x) + C.$$

To determine C we note that $A(x) = 0$ when $x = a$. Substituting these values in the last equation we find

$$0 = F(a) + C,$$

whence $C = -F(a)$ and we have

$$A(x) = F(x) - F(a).$$

To find the area that concerned us at the start we now merely set $x = b$ in the last result, obtaining

$$A = A(b) = F(b) - F(a).$$

We conclude that

$$\int_a^b f(x)\,dx = F(b) - F(a),$$

where $F(x)$ is any particular function having $f(x)$ as its derivative. We denote $F(x) + C$ by the symbolism

$$\int f(x)\,dx,$$

and call this the *indefinite integral* of $f(x)$. Clearly, finding an indefinite integral is the inverse of finding a derivative. It follows that integration and differentiation are inverse operations, just as opening a door and closing a door, filling a tank and emptying a tank, or addition and subtraction, are inverse operations—each operation undoes the other. This is the fundamental theorem of the calculus.

Most present-day calculus texts avail themselves of the fundamental theorem of the calculus by *defining* a definite integral of $f(x)$ as a difference between two evaluations of an associated indefinite integral. From this, then, the original summation definition of a definite integral can be obtained as a theorem. History, of course, proceeded the other way about—a definite integral was defined as the limit of a particular kind of sum, as in Equation (1) above, and it was then later discovered that it can be found as the difference between two evaluations of an indefinite integral.

The fundamental theorem of the calculus not only beautifully relates integration and differentiation as inverse operations, but it permits us to obtain integration rules from differentiation rules. For example, we have that if

$$y = x^{n+1}/(n + 1),$$

where n is a natural number, then

$$dy/dx = x^n,$$

which yields the integration rule

$$\int x^n\,dx = \frac{x^{n+1}}{n + 1} + C.$$

In this way integration becomes, as Augustus De Morgan once remarked, the "memory of differentiation."

In our schools today, subtraction, which is the inverse of addition, is often taught as the "memory of addition." For example, to find $9 - 4$, the pupil asks himself, "4 plus what gives 9?" From recall of his addition tables he knows that it is $4 + 5$ that gives 9, therefore $9 - 4 = 5$.

The fundamental theorem of the calculus can be succinctly stated by the equation

$$\frac{d}{dx} \int_a^x f(x)\, dx = f(x).$$

It follows that the core of any proof of the theorem lies in showing that the problem of finding an area like that pictured in Figure 9 [that is, in finding $F(x) = \int_a^x f(x)\, dx$] reduces to finding a curve [namely, $y = F(x)$] the slope of whose tangent [namely, dF/dx] satisfies a given law [namely, $dF/dx = f(x)$]. The proof that we sketched above is known as the *algebraic proof*, and was first given, at least in its essentials, by both Newton and Leibniz. These two men, however, were not the first to show cognizance of the inverse relation between the operations of integration and differentiation. Indeed, several earlier mathematicians had conjectured the relation and some had established it for certain special situations. Thus Torricelli, about 1646, showed the inverse relation holds for generalized parabolas, that is, for curves with equations of the form $y = x^n$, where n is a natural number. In modern notation, Torricelli showed that

$$\frac{d}{dx} \int_0^x x^n dx = \frac{d}{dx}\left(\frac{x^{n+1}}{n+1}\right) = x^n.$$

James Gregory (1638–1675), the talented Scottish mathematician who died at such a pathetically young age, in his *Geometriae pars universalis* of 1668, may have been the first to see the theorem in its generality. Credit for the first proof, however, is usually given to Isaac Barrow, who, in his *Lectiones opticae et geometricae* of 1669, gave a *geometric proof*. Barrow established the following theorem, here paraphrased (with somewhat different letters) from his work.

THEOREM. *Let ZN be any curve of continually increasing ordinate and* [to render the figure less cluttered] *lying below the axis VM (see Figure 10) and let R be a given line segment. Let VL be a curve such that if an arbitrary ordinate cuts ZN, VM, and VL in F, E, D, respectively, the rectangle of dimensions ED and R has an area equal to that of VEFZ. Finally, let T on VM be such that FE:ED = R:TE. Then TD is the tangent to curve VL at D.*

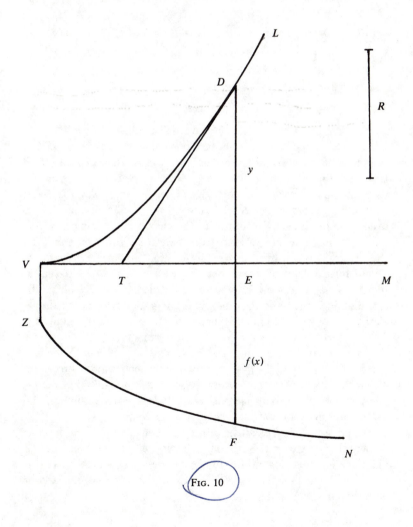

FIG. 10

Barrow furnished a purely geometric proof of this theorem. Assuming the theorem established, taking *VM* as the *x*-axis with origin at *V*, setting $FE = f(x)$ and $ED = y$, and taking *R* as a unit segment, we have

$$y = (ED)(R) = \text{area } VEFZ = \int_0^x f(x)\, dx.$$

We also have

$$dy/dx = \text{slope of curve } VL \text{ at } D = ED/TE = FE/R = f(x).$$

It follows that

$$\frac{d}{dx} \int_0^x f(x)\, dx = f(x),$$

which is the fundamental theorem of the calculus.

Though Barrow's work certainly shows that he was aware of the fundamental theorem of the calculus, his cumbersome geometrical treatment did not permit him to capitalize on the theorem—this had to await the algebraic treatment given later by Newton and Leibniz. With their algebraic approach, Newton and Leibniz were able to utilize the fundamental theorem of the calculus as a means for formally carrying out integrations. Ever since these two men employed the inverse relationship between integration and differentiation to obtain formal rules of integration from formal rules of differentiation, this point of view has persisted in elementary calculus; differentiation is the basic operation and integration is regarded simply as the inverse of this. One has merely to compare this procedure for finding certain areas with the earlier summation procedures for finding the same areas employed by Cavalieri, Torricelli, Roberval, Pascal, Fermat, and Wallis. It accordingly seems quite proper to assign this GREAT MOMENT IN MATHEMATICS to roughly the twenty-five-year period running from Barrow's work of 1669 through the subsequent contributions of Newton and Leibniz.

The impact of the calculus, with integration simplified by the fundamental theorem, was so great, not only on the development of mathematics itself, but on the further progress of civilization, that it has been claimed no one today can consider himself truly educated if he knows nothing of the calculus.

The new tool proved to be almost unbelievably powerful in its astonishingly successful disposal of hosts of problems which had been baffling and quite unassailable in earlier days. Its general methods were able to cope with such matters as lengths of curved arcs, planar areas bounded by quite arbitrary curves, surface areas and volumes of all sorts of solids, intricate maximization and minimization problems, all kinds of problems involving related rates of change, geometrical questions about tangents and normals, asymptotes, envelopes, and curvature, and physical questions about velocity, acceleration, work, energy, power, pressure, centers of gravity, inertia, and gravitational attraction. It was natural that this wide and amazing applicability of the new subject should attract mathematical researchers of the day and that papers should be turned out in great profusion with seemingly little concern regarding the very unsatisfactory foundations of the subject. It was much more exciting to apply the marvelous new tool than to examine its logical soundness, for, after all, the processes employed justified themselves to the researchers in view of the fact that they worked. A conscientious and energetic reëxamination of the logical foundations of the calculus had to outwait a period of frenetic application of the subject. This reëxamination will be considered in a later lecture, for its success constitutes another outstanding GREAT MOMENT IN MATHEMATICS.

As remarked at the conclusion of our previous lecture, the invention of the calculus marks a critical moment in the development of mathematics. It is an instance of the dialectical law of the change of quantity into quality. The slow accumulation of mathematical ideas and material for thought, patiently assembled for a hundred or more years, suddenly, under the genius of Newton and Leibniz, erupted into the invention of a new method and a new point of view, and mathematics was transformed to a higher level. Shelley has compared such moments of great progress in the history of thought to the formation of an avalanche:*

> The sun-awakened avalanche! whose mass,
> Thrice sifted by the storm, had gathered there
> Flake by flake—in heaven-defying minds

---

*This comparison by Shelley has been noted by Alfred North Whitehead in his delightful *An Introduction to Mathematics*, Oxford University Press, 1948.

> As thought by thought is piled, till some great truth
> Is loosened, and the nations echo round,
> ....

Newton expressed the matter somewhat differently when, in generosity to his predecessors, he asserted, "If I have seen farther than others, it is because I have stood upon the shoulders of giants."

Surely no subject in early college mathematics is more exciting or more fun to teach than the calculus. It is like being the ringmaster of a great three-ring circus. It has been said that one can recognize the students on a college campus who have studied the calculus— they are the students with no eyebrows. In utter astonishment at the incredible applicability of the subject, the eyebrows of the calculus students have receded higher and higher and finally vanished over the backs of their heads.

There are other branches of mathematics besides calculus that possess fundamental theorems. Thus there is the *fundamental theorem of algebra* (which asserts that any polynomial equation in one variable with complex coefficients possesses at least one complex root) and the *fundamental theorem of projective geometry* (that says any projective invariant can be expressed in terms of the basic cross-ratio invariant of four collinear points). In a fuller set of lectures, the discovery of these fundamental theorems would also be considered as GREAT MOMENTS IN MATHEMATICS. But in any collection of GREAT MOMENTS IN MATHEMATICS, the discovery of the fundamental theorem of the calculus would surely appear.

Exercises

23.1. What is the inverse of the operation of cubing?

23.2. If u, v, w are functions of x and a is a constant, show that

$$\int (du - dv + dw) = \int du - \int dv + \int dw$$

and

$$\int a\, du = a \int du.$$

23.3. Knowing that $d(\ln x)/dx = 1/x$, find

$$\int \frac{dx}{x}.$$

23.4. Knowing that $d(\sin x)/dx = \cos x$ and $d(\cos x)/dx = -\sin x$, find

$$\int \sin x \, dx \quad \text{and} \quad \int \cos x \, dx.$$

23.5. Find

$$\int \frac{2x - 3}{x^2 - 3x + 5} \, dx.$$

23.6. Using Exercise 23.4 and Leibniz's rule for differentiating a quotient, find $d(\tan x)/dx$. Now find $\int \sec^2 x \, dx$.

23.7. Using Exercise 23.4 find $d(\tan x - x)/dx$. Now find

$$\int \tan^2 x \, dx.$$

23.8. Find the area bounded by the curve $y = 9x - x^3$, the x-axis, and the ordinates at $x = 0$ and $x = 3$.

23.9. Find the area bounded by the curve $y = 2x/(1 + x^2)$, the x-axis, and the ordinates at $x = 0$ and $x = 8$.

23.10. Find the area bounded by the curve $y = 2 \sin x$, the x-axis, and the ordinates at $x = 0$ and $x = \pi/2$.

Further Reading

Boyer, C. B., *The History of the Calculus and Its Conceptual Development.* New York: Dover, 1959.

Child, J. M., *The Geometrical Lectures of Isaac Barrow.* Chicago: Open Court, 1916.

Struik, D. J., *A Source Book in Mathematics, 1200–1800.* Cambridge, Mass.: Harvard University Press, 1969.

POWER(FUL) SERIES

Students of high-school mathematics encounter arithmetic progressions and geometric progressions in their algebra courses. These two progressions are basic and important examples of the more general concept of a *sequence*, and the indicated summation of the terms of either type of progression is an example of the more general concept of a *series*. Let us give formal definitions of these more general concepts.

A *sequence* is a succession of terms, usually formed according to some fixed rule or law, and a *series* is the indicated sum of the terms of a sequence. For example,

$$1, 4, 9, 16, 25$$

and

$$1, -x, x^2/2, -x^3/3, x^4/4, -x^5/5$$

are sequences, while

$$1 + 4 + 9 + 16 + 25$$

and

$$1 - x + x^2/2 - x^3/3 + x^4/4 - x^5/5$$

are their associated series.

The *general*, or *n*th, term of either a sequence or a series is an expression that shows the law of formation of the terms. Thus, in the first example above, the general or *n*th term is n^2. The first term is obtained by setting $n = 1$ and the fourth term by setting $n = 4$. In the second example, the *n*th term is, except for the first term, $(-x)^{n-1}/(n - 1)$.

40

When the number of terms is limited, the sequence or series is
said to be *finite*; when the number of terms is unlimited, the sequence or series is said to be *infinite*. The examples above are illustrations of finite sequences and series. If a sequence or series is
infinite, we indicate it by the first few terms, the nth term, and the
use of dots, as

$$1, 4, 9, \ldots, n^2, \ldots$$

and

$$1 + 4 + 9 + \cdots + n^2 + \cdots.$$

Designating an arbitrary infinite series of constant terms by

$$u_1 + u_2 + u_3 + \cdots + u_n + \cdots,$$

we say that the series *converges* if

$$\lim_{n \to \infty} (u_1 + u_2 + \cdots + u_n)$$

exists and is finite; otherwise we say the infinite series *diverges*. In
the former case, if the limit is the number S, we say that S is the *sum*
of the infinite series, and we write

$$S = u_1 + u_2 + \cdots + u_n + \cdots = \sum_{n=1}^{\infty} u_n.$$

In elementary calculus courses the student learns several tests for
convergence or divergence of infinite series. We cannot go into these
tests in our brief lecture, but the student who has studied infinite
series will recall the beauty and cleverness of the tests, and the
remarkable applications that can be made of convergent infinite
series.

The high-school student will recall that an *arithmetic progression*
is a sequence of the form

$$a, a + d, a + 2d, \ldots, a + (n - 1)d, \ldots,$$

where a and d are fixed constants, and a *geometric progression* is a
sequence of the form

$$a, ar, ar^2, \ldots, ar^{n-1}, \ldots,$$

where a and r are fixed constants. These are the oldest kinds of sequences to have been considered and instances of them can be found in the mathematics of the ancient Egyptians and Babylonians of two millennia B.C. The first person, however, to sum a convergent infinite series was Archimedes, who, in his treatise *Quadrature of the Parabola*, showed that the geometric series

$$1 + \frac{1}{4} + \frac{1}{4^2} + \cdots + \frac{1}{4^{n-1}} + \cdots$$

has the sum 4/3. It is interesting to pause for a moment to see just how this geometric series arose in Archimedes' work. Let C, D, E be points on the arc of the parabolic segment shown in Figure 11, these points being obtained by drawing LC, MD, NE, parallel to the axis of the parabola, through the midpoints L, M, N of AB, CA, CB. From the geometry of the parabola Archimedes showed that

$$\Delta CDA + \Delta CEB = \frac{\Delta ACB}{4}.$$

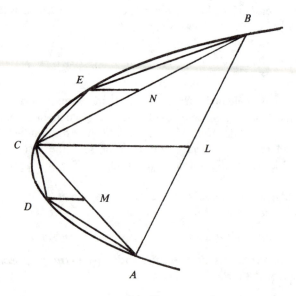

FIG. 11

By repeated applications of this idea it follows that the area of the parabolic segment is given by

$$\Delta ABC + \frac{\Delta ABC}{4} + \frac{\Delta ABC}{4^2} + \cdots + \frac{\Delta ABC}{4^{n-1}} + \cdots$$

$$= \Delta ABC\left(1 + \frac{1}{4} + \frac{1}{4^2} + \cdots + \frac{1}{4^{n-1}} + \cdots\right).$$

To find the sum of the geometric series Archimedes employed the double *reductio ad adsurdum* apparatus of the ancient method of exhaustion.

Though Archimedes was the first to sum a particular convergent infinite series, the first person adequately to consider the concept of convergence of an infinite series appears to have been the eminent German mathematician Carl Friedrich Gauss (1777-1855), in a treatment of hypergeometric series made by him in 1812.*

This pioneering work of Archimedes and Gauss are certainly important episodes in the history of infinite series. For our GREAT MOMENTS IN MATHEMATICS in this area of mathematics, however, we choose two other important episodes. The first will constitute the principal subject matter of the present lecture and will be concerned with the work done by Brook Taylor in 1715 and Colin Maclaurin in 1742 in connection with power series. The other will be considered in our next lecture and will be concerned with the work done by Joseph Fourier in 1807-1822 in connection with trigonometric series.

An infinite series of the form

$$a_0 + a_1x + a_2x^2 + \cdots + a_{n-1}x^{n-1} + \cdots,$$

---

*A *hypergeometric series* is a series of the form

$$1 + \frac{a}{1} \cdot \frac{b}{c}x + \frac{a(a+1)}{2!} \cdot \frac{b(b+1)}{c(c+1)}x^2$$

$$+ \frac{a(a+1)(a+2)}{3!} \cdot \frac{b(b+1)(b+2)}{c(c+1)(c+2)}x^3 + \cdots,$$

where a, b, c are any real numbers different from 0, -1, $-2, \ldots$.

involving positive integral powers of a variable x with constant coefficients a_i, is called a *power series in x*. It is a generalization of the familiar polynomial

$$a_0 + a_1x + \cdots + a_nx^n.$$

More generally, an infinite series of the form

$$a_0 + a_1(x - a) + a_2(x - a)^2 + \cdots + a_{n-1}(x - a)^{n-1} + \cdots$$

is called a *power series in x − a*.

If, in the power series in x above, a particular value is assigned to x, the series reduces to a series of constant terms, which may or may not converge. It is obvious that the power series converges for the value $x = 0$. It might converge for no other value of x, or, again, it might converge for some other values of x, maybe for all values of x. In elementary calculus texts it is shown that if the power series converges for $x = b$, where b is a positive constant, then it will converge for all values of x such that $|x| < b$, and if the power series diverges for $x = b$, then it will diverge when $|x| > b$. It follows that the totality of values of x for which the series converges constitutes a number interval of the form $-b < x < b$, perhaps along with one or both of the end-values of the interval. The interval $-b < x < b$ (or $-\infty < x < \infty$, if the series should converge for all values of x) is known as the *interval of convergence* of the power series; it is centered at $x = 0$. Similarly, a power series in $x - a$ possesses an interval of convergence of the form

$$a - b < x < a + b,$$

centered at $x = a$.

A convergent power series in x is obviously a function of x for all values of x in its interval of convergence. Thus we may write

$$f(x) = a_0 + a_1x + \cdots + a_{n-1}x^{n-1} + \cdots.$$

When a known function is written in the form of a power series in x, the function is said to be *expanded in a power series in x*. For example, it can be shown that the geometric series

$$1 + x + x^2 + \cdots + x^{n-1} + \cdots$$

converges to $1/(1 - x)$ for all values of x in the interval of convergence $-1 < x < 1$. Therefore,

$$\frac{1}{1 - x} = 1 + x + x^2 + \cdots + x^{n-1} + \cdots, \qquad -1 < x < 1.$$

If a given function $f(x)$ is represented by a power series in x, it is natural to wonder what must be the form of the coefficients a_0, $a_1, \ldots, a_{n-1}, \ldots$. To answer this we might proceed as follows. In

$$f(x) = a_0 + a_1 x + a_2 x^2 + \cdots + a_{n-1} x^{n-1} + \cdots$$

set $x = 0$. We find that we must have

$$f(0) = a_0.$$

Now *assume* that the series may be differentiated term by term, and that this differentiation can be continued. Then we shall have

$$f'(x) = a_1 + 2a_2 x + 3a_3 x^2 + \cdots + (n - 1)a_{n-1} x^{n-2} + \cdots,$$
$$f''(x) = 2a_2 + 6a_3 x + \cdots + (n - 1)(n - 2)a_{n-1} x^{n-3} + \cdots,$$
$$f'''(x) = 6a_3 + \cdots + (n - 1)(n - 2)(n - 3)a_{n-1} x^{n-4} + \cdots,$$

and so on. Setting $x = 0$ we find, employing the factorial symbol,*

$$f'(0) = a_1, f''(0) = 2!a_2, f'''(0) = 3!a_3, \ldots,$$
$$f^{(n-1)}(0) = (n - 1)!a_{n-1}, \ldots.$$

Solving for $a_1, a_2, \ldots, a_{n-1}, \ldots$, and substituting in the original series representation, we find

$$f(x) = f(0) + \frac{f'(0)}{1!} x + \frac{f''(0)}{2!} x^2 + \cdots + \frac{f^{(n-1)}(0)}{(n - 1)!} x^{n-1} + \cdots.$$

It can similarly be shown, more generally, that if a given infinitely differentiable function $f(x)$ can be represented by a power series in $x - a$, then, within the interval of convergence of the series, we have

---

*By *factorial r,* indicated by $r!$, where r is a positive integer, is meant the product $(1)(2)(3) \cdots (r - 1)(r)$.

$$f(x) = f(a) + f'(a)(x - a) + \cdots + \frac{f^{(n-1)}(a)}{(n - 1)!}(x - a)^{n-1} + \cdots.$$

The expansion of $f(x)$ as a power series in $x - a$ appeared in 1715, with no consideration of convergence, in the *Methodus incrementorum directa et inversa* of the Englishman Brook Taylor (1685–1731), and has since become known as the *Taylor expansion* of $f(x)$ about $x = a$. The expansion of $f(x)$ as a power series in x, which is nothing but the Taylor expansion for the case $a = 0$, has become known as the *Maclaurin expansion* of $f(x)$, though this particular case was explicitly given by Taylor in his work and by James Stirling (1692–1770) a couple of years later. It became known as the Maclaurin expansion when it was used in 1742, with acknowledgment to both Taylor and Stirling, in the great and influential two-volume *Treatise of Fluxions* of the Scotsman Colin Maclaurin (1698–1746).

Taylor applied his series to the solution of numerical equations as follows. Let a be an approximation to a root of $f(x) = 0$. Set $f(a) = k$, $f'(a) = k'$, $f''(a) = k''$, and $x - a = h$. Expand $f(x)$ into its Taylor expansion about $x = a$. Discard all powers of h above the second. Substitute the values of k, k', k'' and solve for h. Then $a + h$ is a better approximation of the required root. By successive applications of this process, closer and closer approximations can be obtained. Some work done by Taylor in the theory of perspective found application in modern times in the mathematical treatment of photogrammetry, the science of surveying by means of photographs taken from the underside of an airplane.

Recognition of the full importance of Taylor's series was delayed until 1755, when Euler brilliantly applied them in his differential calculus, and, still later, when Lagrange in 1797 used the series as the foundation of his theory of functions.

Maclaurin was one of the ablest mathematicians of the eighteenth century. He did very notable work in geometry, particularly in the study of higher plane curves, and he showed great power in applying classical geometry to physical problems. Among his many papers in applied mathematics is a prize-winning memoir on the mathematical theory of tides. In his *Treatise of Fluxions* appears his investigation of the mutual attraction of two ellipsoids of revolution.

Maclaurin was a mathematical prodigy. He matriculated at the

University of Glasgow at the age of 11. At 15 he took his master's degree and gave a remarkable public defense of his thesis on gravitational attraction. At 19 he was elected to the chair of mathematics at the Marischal College in Aberdeen, and at 21 published his first important work, *Geometria organica*. At 27 he became deputy, or assistant, to the professor of mathematics at the University of Edinburgh. There was some difficulty in obtaining a salary to cover his assistantship, and Newton offered to bear the cost personally so that the university could secure the services of so outstanding a young man. In time Maclaurin succeeded the man he assisted. His treatise on fluxions appeared when he was 44, only four years before he died; this was the first logical and systematic exposition of Newton's method of fluxions and was written by Maclaurin as a reply to attacks made by Bishop George Berkeley on the principles of the calculus.

It is only when one studies the calculus that one becomes cognizant of the great usefulness of the Taylor and Maclaurin expansions. Certainly every precalculus mathematics student must, at one time or another, have wondered how the extensive tables of trigonometric, logarithmic, and exponential functions were obtained. The answer is that they were largely computed with the aid of power series.

Suppose, for example, one wishes to compute, approximately, the value of e. The reader can readily show that the Maclaurin expansion of e^x, which converges for all x, is

$$e^x = 1 + x + x^2/2! + x^3/3! + \cdots + x^{n-1}/(n-1)! + \cdots.$$

If $x = 1$, this becomes

$$e = 1 + 1 + 1/2! + 1/3! + 1/4! + \cdots$$
$$= 1 + 1 + 0.5 + 0.166667 + 0.041667$$
$$+ 0.008333 + 0.001389 + 0.000198 + \cdots.*$$

It follows that, approximately, $e = 2.718254$, which is correct to four decimal places.

Again, suppose we wish to compute $\sin 10° = \sin (\pi/18)$ correct

---

*Note that these terms are easily calculated inasmuch as the ith term is merely the $(i - 1)$th term divided by i.

to five decimal places. The Maclaurin expansion of sin x, which converges for all x, is readily found to be

$$\sin x = x - x^3/3! + x^5/5! - x^7/7! \quad \cdots .$$

It follows that

$$\sin 10° = \frac{\pi}{18} - \frac{1}{3!}\left(\frac{\pi}{18}\right)^3 + \frac{1}{5!}\left(\frac{\pi}{18}\right)^5 - \cdots .$$

This is an alternating series (that is, the terms are alternately positive and negative). But it is known that the error incurred in using only a few terms of an alternating series cannot exceed the numerical value of the first term omitted. It follows that

$$\sin 10° = 0.174532 - 0.000886 + 0.000001 - \cdots .$$

The third term does not affect the fifth decimal place. Therefore, taking only the first two terms in the expansion, we find, correct to five decimal places,

$$\sin 10° = 0.17365.$$

Once more, let us use power series to approximate the value of

$$\int_0^1 \sin x^2 \, dx.$$

Set $z = x^2$. Now, by a Maclaurin expansion,

$$\sin z = z - z^3/3! + z^5/5! - \cdots ,$$

whence

$$\sin x^2 = x^2 - x^6/3! + x^{10}/5! - \cdots ,$$

and, since one can integrate a convergent power series term by term between any two limits within the interval of convergence of the series,

$$\int_0^1 \sin x^2 \, dx = \int_0^1 \left(x^2 - \frac{x^6}{3!} + \frac{x^{10}}{5!}\right) dx, \text{ approximately,}$$

$$= \left[\frac{x^3}{3} - \frac{x^7}{42} + \frac{x^{11}}{1320} \right]_0^1$$

$$= 0.3333 - 0.0238 + 0.0008$$

$$= 0.3103.$$

The fantastic computation of π to over a half-million decimal places, which was performed in 1967, was done with the aid of power series and modern computing machinery. The student of power series has many beautiful and exciting times in store for him. The discovery of the Taylor and Maclaurin expansions certainly marks a GREAT MOMENT IN MATHEMATICS.

Exercises

24.1. (a) Find a formula for the sum of the first n terms of an arithmetic progression having a for its first term and d for its common difference.

(b) Find a formula for the sum of the first n terms of a geometric progression having a for its first term and r for its common ratio.

(c) If $|r| < 1$, shows that

$$\sum_{n=0}^{\infty} ar^n = \frac{a}{1-r}.$$

24.2. (a) Solve the following problem found in the ancient Egyption Rhind papyrus (ca. 1650 B.C.): "Divide 100 loaves among five men in such a way that the shares received shall be in arithmetic progression and that one seventh of the sum of the largest three shares shall be equal to the sum of the smallest two."

(b) Otto Neugebauer (b. 1899) has found some interesting series problems on an ancient Babylonian tablet dating from about 300 B.C. One of these states that

$$1 + 2 + 2^2 + \cdots + 2^9 = 2^9 + 2^9 - 1.$$

Verify this summation.

(c) Verify Archimedes' summation

$$1 + 1/4 + 1/4^2 + \cdots + 1/4^{n-1} + \cdots = 4/3.$$

24.3. (a) Show, by mathematical induction, that

$$\sum_{i=1}^{n} i^2 = \frac{n(n+1)(2n+1)}{6} = \frac{2n+1}{3} \sum_{i=1}^{n} i.$$

(b) Otto Neugebauer has found an ancient Babylonian tablet dating from about 300 B.C. stating that

$$1^2 + 2^2 + 3^2 + \cdots + 10^2 = \left[1\left(\frac{1}{3}\right) + 10\left(\frac{2}{3}\right)\right]55 = 385.$$

Show how this might be an application of the formula of Exercise 24.3 (a).

24.4. Paralleling the treatment given in the lecture for finding the coefficients of the expansion of a given function $f(x)$ as a power series in x, find the coefficients of the expansion of $f(x)$ as a power series in $x - a$.

24.5. (a) Find the Maclaurin expansions of e^x, $\sin x$, and $\cos x$.
(b) Assuming the Maclaurin expansion of e^x holds for complex x, formally show (as did Euler) that

$$e^{ix} = \cos x + i \sin x.$$

(c) From Exercise 24.5 (b) obtain the relation

$$e^{i\pi} + 1 = 0,$$

connecting five of the most important numbers in mathematics.

24.6. Compute $\sqrt{e}$ correct to four decimal places.

24.7. Using the Taylor expansion of $\cos x$ about $x = \pi/4$, find $\cos 44°$ correct to five decimal places.

24.8. Find the Maclaurin expansion for $\cos x$ by differentiating the Maclaurin expansion for $\sin x$.

24.9. Evaluate $\displaystyle\int_0^1 \cos\sqrt{x}\ dx$ to three decimal places.

24.10. Evaluate $\displaystyle\int_0^1 e^{-x^2}\ dx$ to three decimal places.

Further Reading

HEATH, T. L., *The Works of Archimedes*. New York: Cambridge University Press, 1897. Reprinted by Dover.

KNOPP, KONRAD. *Theory and Application of Infinite Series,* transl. by R. C. Young. London: Blackie, 1928.

YEA + YEA + YEA + YEA

The seventeenth century was a spectacular period in the development of mathematics. Early in the century Napier published his invention of logarithms; Harriot and Oughtred contributed noteworthily to the notation and systematization of algebra; Galileo founded the science of dynamics; and Kepler announced his three famous laws of planetary motion. Later in the century Desargues and Pascal opened the new field of projective geometry; Descartes launched modern analytic geometry; Fermat laid the foundations of modern number theory; Pascal, Fermat, and Huygens created the field of mathematical probability; and the first computing machines were invented by Pascal and Leibniz. Then, toward the end of the century, after a number of mathematicians had prepared the way, the epoch-making creation of the calculus was made by Newton and Leibniz. Thus, during the seventeenth century, many new and vast mathematical fields were opened up for investigation. The dawn of modern mathematics was at hand.

The calculus, assisted by analytic geometry, was certainly the greatest mathematical tool discovered in the seventeenth century. Its remarkable applicability and its highly successful disposal of a wide variety of problems that were quite unapproachable in precalculus days attracted such a large number of mathematical researchers that it can be said, with fair honesty, that much of the mathematics of the eighteenth century and the early part of the nineteenth century was devoted to exploiting the new and powerful tool. It is true that often, in the resulting pell-mell application of the discipline, too little regard was given to the unsatisfactory logical foundations of the subject. At the time, the idea of function itself was only poorly understood, and such notions as limit, continuity, differentiability, integrability, and convergence were very hazy. Rigorous clarification of these basic concepts had to await the nineteenth-century critical reëxamination of the foundations of the calculus.

Nevertheless, a great deal was accomplished during the more than a century of intuitive application of the calculus. Some of the great masters of applied calculus of this period were Jacob Bernoulli (1654–1705), Johann Bernoulli (1667–1748), Leonhard Euler (1707–1783), Claude Alexis Clairaut (1713–1765), Jean-le-Rond d'Alembert (1717–1783), Johann Heinrich Lambert (1728–1777), Joseph Louis Lagrange (1736–1813), Pierre-Simon Laplace (1749–1827), and Adrien-Marie Lengendre (1752–1833).

Among the accomplishments and masterpieces of these men are: (1) Jacob Bernoulli's study of the catenary curve with extensions to strings of variable density and strings under the action of a central force, the discovery of the so-called *isochrone*, or curve along which a body will fall with uniform vertical velocity, the form taken by an elastic rod fixed at one end and carrying a weight at the other, the form assumed by a flexible rectangular sheet having two opposite edges held horizontally fixed at the same height and loaded with a heavy liquid, and the shape of a rectangular sail filled with wind; (2) Johann Bernoulli's work on optical phenomena connected with reflection and refraction, the determination of the orthogonal trajectories of families of curves, and his contribution to the problem of the *brachystochrone*, the curve of quickest descent of a weighted particle moving between two points in a gravitational field; (3) Euler's prestigious two-volume *Introductio in analysin infinitorum* of 1748, his rich *Institutiones calculi differentialis* of 1755, the allied three-volume *Institutiones calculi integralis* of 1768–74, and others on mechanics, which served as textbooks and models of textbooks for many years, plus his countless papers on applied subjects; (4) Clairaut's definitive work *Théorie de la figure de la Terre* (1743) and his prize-winning paper *Théorie de la Lune* (1752); (5) d'Alembert's *Traité de dynamique* (1743), based upon the great principle of kinetics that now bears his name, and his subsequent treatises on the equilibrium and motion of fluids (1744), the causes of winds (1746), and vibrating strings (1757); (6) Lambert's development of hyperbolic functions and his work on the determination of comet orbits; (7) Lagrange's monumental *Mécanique analytique* (1788) and his great publication *Théorie des fonctions analytiques contenant les principes du calcul différentiel* (1797); (8) Laplace's cyclopean five-volume *Traité de mécanique céleste* (1799–1825), which earned him the title of "the Newton of France," and his *Théorie analytique des probabilités* (1812); (9) Legendre's work on elliptic functions, the method of least squares, and differential equations.

Many of these works can well vie for an honored position among the GREAT MOMENTS IN MATHEMATICS. For lack of space and time we pass over them and now nominate a memorable session of the French Academy of Sciences that was held on the 21st of December in the year 1807. At that gathering the thirty-nine-year-old mathematician and engineer Joseph Fourier proclaimed a thesis that initiated a new and highly fruitful chapter in the history of mathematics.

Fourier, like a number of scientists of his time, had become interested in the practical problem of the flow of heat in metalic rods, plates, and solid bodies, and had submitted for presentation before the French Academy a basic paper on heat conduction. During his presentation of the paper, he made the startling claim that any function defined in a finite closed interval by any arbitrarily drawn graph can be resolved into a sum of pure sine and cosine functions. To be more explicit, Fourier claimed that *any* function whatever, no matter how capriciously it is defined in the interval $(-\pi, \pi)$, can be represented in that interval by

$$\frac{a_0}{2} + \sum_{n=1}^{\infty} (a_n \cos nx + b_n \sin nx), \tag{1}$$

where the a's and the b's are suitable real numbers. Such a series is known as a *trigonometric series* and was not new to the mathematicians of the time. Indeed, a number of more or less well behaved functions had been shown to be representable by such a series. But Fourier claimed that *any* function defined in $(-\pi, \pi)$ can be so represented.

The situation is illustrated in Figure 12, showing a random graph in the interval $(-\pi, \pi)$. Since the sine and cosine functions are periodic with period 2π, it follows that any function represented by a trigonometric series is also periodic with period 2π, and in Figure 12 the graph would be repeated over and over to the left of $-\pi$ and to

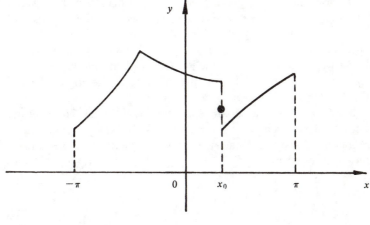

FIG. 12

the right of $+\pi$. It follows that one does not have to choose the interval $(-\pi, \pi)$ as the period interval—any interval $(c, c + 2\pi)$ of length 2π would serve just as well.

The academicians at the session were very skeptical of Fourier's claim. Since the sine and cosine functions are analytic (that is, are infinitely differentiable), then, surely, they said, any sum of such functions must also be analytic, and analytic functions are far indeed from being arbitrarily capricious in a given interval. An analytic function, for instance, has the property that if its graph is known in any arbitrarily small interval, such as just a tiny piece of $(-\pi, \pi)$, then its graph is uniquely determined everywhere else. How can this fact be reconciled with the arbitrariness, claimed by Fourier, of the graph throughout the interval $(-\pi, \pi)$?

The paper, which was judged by Lagrange, Laplace, and Legendre, was accordingly rejected, but, to encourage Fourier to develop his ideas more carefully, the French Academy made the problem of heat propagation the subject of a grand prize to be awarded in 1812. Fourier submitted a revised paper in 1811, which was judged by a group containing, among others, the former three judges, and the paper won the prize, though it was criticized for lack of rigor and so was not recommended for publication in the Academy's *Mémoires*.

Resentful, Fourier continued his researches on heat, and in 1822 published one of the great classics of mathematics, his *Théorie analytique de la chaleur* (*The Analytical Theory of Heat*). This work incorporated the first part of his 1811 paper essentially unchanged and became the main source for Fourier's ideas. Two years later, Fourier became secretary of the French Academy and in that capacity was able to have his 1811 paper published in its original form in the Academy's *Mémoires*.

It can be shown, assuming that the series (1) can be integrated term by term from $-\pi$ to $+\pi$, that *if* a function $f(x)$ can be represented by a trigonometric series (1), then the coefficients in the series are given by

$$a_n = \frac{1}{\pi} \int_{-\pi}^{\pi} f(x) \cos nx \, dx,$$
$$b_n = \frac{1}{\pi} \int_{-\pi}^{\pi} f(x) \sin nx \, dx. \qquad (n \geq 0) \qquad (2)$$

These formulas had been known to Euler, but they became more generally known through the work of Fourier; and the expressions are today called the *Fourier coefficients* of $f(x)$.

Calling the trigonometric series (1), with its coefficients determined by (2), the *Fourier series* of $f(x)$, the following question naturally arises. For what $f(x)$ does its Fourier series actually converge to the values of $f(x)$ in $(-\pi, \pi)$? Fourier claimed that the series does so for *any* $f(x)$, but he was able to support his claim with only some very loose arguments. It turns out that, though his series represents $f(x)$ for a very broad class of functions $f(x)$, his claim that it does so for *any* function is too extravagant. Later, in 1829, the German mathematician Peter Gustav Lejeune Dirichlet (1805–1859) proved the following important theorem.

DIRICHLET'S THEOREM. *If in the closed interval $[-\pi, \pi]$ $f(x)$ is single-valued and bounded, has only a finite number of discontinuities, and has only a finite number of maxima and minima, then the Fourier series of $f(x)$ converges to $f(x)$ at all points where $f(x)$ is continuous and to the average of the right-hand and left-hand limits of $f(x)$ at each point where $f(x)$ is discontinuous.*

Figure 12 shows the circumstances of Dirichlet's Theorem at a point x_0 of discontinuity; the value of the series at x_0 is given by the ordinate of the midpoint of the "jump" in the graph at that point.

Dirichlet's conditions are *sufficient,* but not *necessary,* conditions for convergence. To this day no conditions that are both necessary and sufficient have been found, but sufficient conditions broader than Dirichlet's have been achieved. Nevertheless, Dirichlet's conditions hold for practically all periodic functions encountered in physics and engineering.

The largely arbitrary nature of the functions representable by Fourier series renders these series a far more powerful instrument in higher analysis than the much more restrictive power series considered in our previous lecture. The Fourier series have proved to be highly valuable in such fields of study as acoustics, optics, electrodynamics, thermodynamics, and many others, and they play a cardinal role in harmonic analysis, beam and bridge problems, and in the solution of differential equations. Indeed, it was the Fourier series that motivated the modern methods in mathematical physics involving the integration of partial differential equations subject to boundary conditions. In a later lecture we shall see the important role played by the series in the evolution of the function concept.

Time has vindicated Fourier far more than it has his original critics. Though Fourier's claim about his series was perhaps too great, the restrictions of his critics were much too confining. The critics committed the error, quite common at the time, of assuming that a property holding for the terms of a convergent series, such as continuity or differentiability, must also hold for the sum of the series.

Jean-Baptiste-Joseph Fourier was born in Auxerre in 1768 and died in Paris in 1830. Orphaned at the age of eight, he was educated in the local military school conducted by the Benedictines. As the son of a tailor he was denied a commission in the army, but, because of his high scholarship, he was given a lectureship in mathematics at the military school. Following the portentous year of 1789, he joined the people's party and with great enthusiasm threw himself behind the promotion of the French Revolution and was rewarded by a chair at the École Polytechnique. He resigned from this position so that he, along with the mathematician Gaspard Monge, could accompany Napoleon on the latter's Egyptian campaign. In 1798 he was appointed governor of Lower Egypt. Following the British victories and the capitulation of the French in 1801, Fourier returned to France and was made prefect of Grenoble. It was while at Grenoble that he commenced his experiments on heat. In 1816 he moved to Paris.

In addition to his great treatise on the theory of heat published in 1822, an unfinished work by Fourier was edited and posthumously published in 1831; it contains, among other original matters, Fourier's work, considered in present-day texts on the theory of equations, on the position of the roots of a polynomial equation.

In most engineering applications of trigonometric series, one requires an expansion of a given function over an interval of length different from 2π, say of length $2L$. We can regard the interval $(-L, L)$ as the result of elongating (or compressing) the interval $(-\pi, \pi)$ in the ratio L/π. Thus, if we denote by z the variable referred to the latter interval, we must have $x/z = L/\pi$, or $x = Lz/\pi$. Now suppose we represent $f(x) = f(Lz/\pi)$, regarded as a function of z, by a Fourier series valid for $-\pi < z < \pi$. Substituting $z = \pi x/L$ in that series, we obtain a representation of $f(x)$ valid for $-L < x < L$. Fourier conceived the idea of letting L, in this latter series, become infinite, and thus arrived at his most remarkable and original creation, the so-called *Fourier integral* as the limit of a Fourier series when the period of the involved function becomes infinite. We cannot go into this here, but all students of advanced mathematics who study this material express the immeasurable pleasure derived therefrom.

Lord Kelvin (William Thomson, 1824–1907) maintained that his whole career in mathematical physics was influenced by Fourier's work on heat, and Clerk Maxwell (1831–1879) pronounced Fourier's treatise "a great mathematical poem."

An amusing story is told about Fourier and his interest in heat. It seems that from his experience in Egypt, and maybe his work on heat, he became convinced that desert heat is the ideal condition for good health. He accordingly clothed himself in many layers of garments and lived in rooms of unbearably high temperature. It has been said that this obsession with heat hastened his death, by heart disease, so that he died, thoroughly cooked, in his sixty-third year.

Perhaps Fourier's most quoted remark (it appeared in his early work on the mathematical theory of heat) is: "The deep study of nature is the most fruitful source of mathematical discovery."

Let us now close the lecture with a few simple illustrations of Fourier series, leaving details of calculation as exercises.

EXAMPLE 1. Consider the function $f(x)$ defined by the relations (see Figure 13)

$$f(x) = 2, \qquad -\pi < x < 0,$$
$$f(x) = 1, \qquad 0 < x < \pi.$$

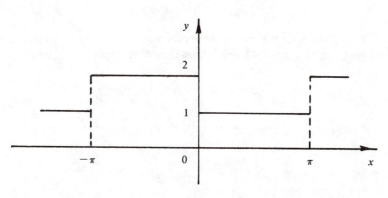

FIG. 13

We have not defined $f(x)$ at $x = 0$, but Dirichlet's Theorem indicates that the Fourier series of $f(x)$ will yield $f(0) = 3/2$. Moreover, the series, because of its periodic character, will define $f(x)$ for every x not in the interval $(-\pi, \pi)$, and we should expect to get $f(-\pi) = f(\pi) = 3/2$. The Fourier series of $f(x)$ is easily found to be

$$f(x) = \frac{3}{2} - \frac{2}{\pi}\left(\sin x + \frac{1}{3}\sin 3x + \frac{1}{5}\sin 5x + \cdots\right).$$

Note that $f(-\pi) = f(0) = f(\pi) = 3/2$, as predicted. Setting $x = \pi/2$, we obtain, since $f(\pi/2) = 1$,

$$1 = \frac{3}{2} - \frac{2}{\pi}\left(1 - \frac{1}{3} + \frac{1}{5} - \frac{1}{7} + \cdots\right)$$

or

$$\frac{\pi}{4} = 1 - \frac{1}{3} + \frac{1}{5} - \frac{1}{7} + \cdots,$$

an alternating convergent series representing the ratio $\pi/4$.

EXAMPLE 2. Consider the function $f(x)$ defined by the relations (see Figure 14)

$$f(x) = -x, \qquad -\pi < x \le 0,$$
$$f(x) = \quad x, \qquad 0 < x < \pi.$$

Though this function has no discontinuities in $(-\pi, \pi)$, it is nondifferentiable at $x = 0$. The Fourier series of the function is

$$f(x) = \frac{\pi}{2} - \frac{4}{\pi}\left(\cos x + \frac{1}{3^2}\cos 3x + \frac{1}{5^2}\cos 5x + \cdots\right).$$

Substituting $x = 0$, we obtain another interesting series involving π,

$$0 = \frac{\pi}{2} - \frac{4}{\pi}\left(1 + \frac{1}{3^2} + \frac{1}{5^2} + \frac{1}{7^2} + \cdots\right).$$

or

$$\frac{\pi^2}{8} = 1 + \frac{1}{3^2} + \frac{1}{5^2} + \frac{1}{7^2} + \cdots.$$

EXAMPLE 3. Consider the function $f(x)$ defined by (see Figure 15)

$$f(x) = x, \qquad -\pi < x < \pi.$$

This function coincides with that of Example 2 in the right-hand half of the interval $(-\pi, \pi)$, but differs in the left-hand half. The Fourier series here is

$$f(x) = 2\left(\sin x - \frac{1}{2}\sin 2x + \frac{1}{3}\sin 3x - \cdots\right).$$

Note that $f(-\pi) = f(\pi) = 0$, agreeing with the values predicted by Dirichlet's Theorem.

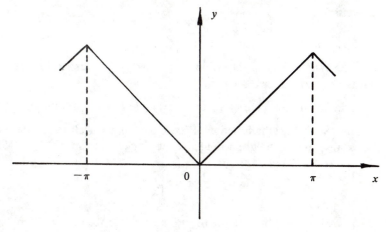

FIG. 14

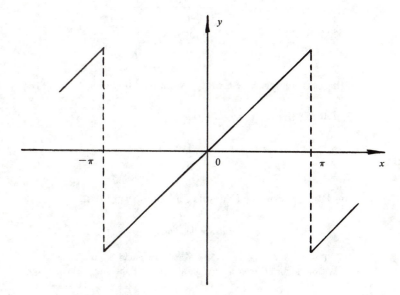

FIG. 15

Exercises

25.1. Show that $\int_{-\pi}^{\pi} \sin nx\, dx = \int_{-\pi}^{\pi} \cos nx\, dx = 0$ when $n \neq 0$.

25.2. Why is the constant term in the series (1) of the lecture text written as $a_0/2$ rather than as a_0?

25.3. Obtain the Fourier series of Example 1 of the lecture text.

25.4. Show that the function of Example 1 of the lecture text can be expressed over the given range by the single equation

$$f(x) = \frac{3}{2} - \frac{x}{2|x|}.$$

25.5. Show that the sum of the squares of the reciprocals of the positive zeros of $\cos x$ is equal to $1/2$.

25.6. Using the integration formulas

$$\int x \cos nx\, dx = \frac{x \sin nx}{n} + \frac{\cos nx}{n^2} + C,$$

$$\int x \sin nx\, dx = -\frac{x \cos nx}{n} + \frac{\sin nx}{n^2} + C,$$

obtain the Fourier series of Examples 2 and 3 of the lecture text.

25.7. Obtain the convergent series

$$\frac{\pi}{4} = 1 - \frac{1}{3} + \frac{1}{5} - \frac{1}{7} + \cdots$$

from the Fourier series of Example 3 of the lecture text.

Further Reading

BELL, E. T., *Men of Mathematics*. New York: Simon and Schuster, 1937.
LANCZOS, CORNELIUS, *Discourse on Fourier Series*. Edinburgh: Oliver & Boyd 1966.

THE LIBERATION OF GEOMETRY, I

We shall now recount, in turn, two very remarkable and revolutionary mathematical developments that occurred in the first half of the nineteenth century. The first one was the discovery, about 1829, of a self-consistent geometry markedly different from the familiar geometry of Euclid; the second was the discovery, in 1843, of an algebra radically different from the customary algebra of the real number system. Each of these developments merits inclusion in any selection of GREAT MOMENTS IN MATHEMATICS, and to do them justice we shall devote two lectures to each one. In each case our treatment will be strongly historical, for each development is concerned with fundamental mathematical ideas, and a genuine understanding of ideas is not possible without an analysis of origins.

The first of the two developments constitutes one of the great sagas in the history of human thought. Like a number of the thrilling stories in the evolution of mathematics, one finds its beginnings back in the glorious days of the ancient Greeks, for it arose from criticisms of the fifth, or parallel, postulate of Euclid's *Elements*:

> If a straight line falling on two straight lines makes the interior angles on the same side together less than two right angles, the two straight lines, if produced indefinitely, meet on that side on which the angles are together less than two right angles.

---

*The material for LECTURE 26 and LECTURE 27 has been adapted from the fuller treatment given in Chapter 3 of H. Eves and C. V. Newsom, *An Introduction to the Foundations and Fundamental Concepts of Mathematics*, revised edition, Holt, Rinehart and Winston, 1965. This latter material, in turn, was expanded from the earlier unpublished notes of the lecture series on GREAT MOMENTS IN MATHEMATICS.

The fifth-century neo-Platonic philosopher and commentator Proclus tells us, in his *Commentary on Euclid, Book I,* that this postulate was attacked from the very start. Even a cursory reading of Euclid's five postulates* immediately discloses a very noticeable difference between the fifth postulate and the other four; the fifth postulate is verbose and involved, lacking the terseness and easy comprehensibility possessed by the other four, and it certainly does not have the quality of "self-evidence" or ready acceptance demanded by the material axiomatics of the Greeks.† A more studied examination reveals that the fifth postulate is actually the converse of Euclid's Proposition I 17. It is not surprising that to many it seemed more like a proposition than a postulate. Moreover, Euclid himself seems to have employed it with reluctance, for he defers its use until he reaches Proposition I 29.

Now if one is unhappy with a particular postulate in an axiomatic development of some field of study, there is a choice of only two things that one might do about it—either replace the postulate by a more acceptable equivalent, or do away with the postulate altogether by deriving it as a theorem from the other postulates of the field of study. Attempts along each of these lines were made from the earliest times.

Some of the alternatives for the parallel postulate that have been either proposed or tacitly assumed over the years are:

(1) *There exists a pair of coplanar straight lines everywhere equally distant from one another.*

(2) *Through a given point not on a given straight line can be drawn only one straight line parallel to the given line.*

(3) *There exists a pair of similar noncongruent triangles.*

(4) *If in a quadrilateral a pair of opposite sides are equal and if the angles adjacent to a third side are right angles, then the other two angles are also right angles.*

(5) *If in a quadrilateral three angles are right angles, then the fourth angle is also a right angle.*

(6) *There exists at least one triangle having the sum of its three angles equal to two right angles.*

---

*For statements of these postulates, see LECTURE 8 or the Appendix to LECTURE 27.

†For a review of material axiomatics, see LECTURE 7.

(7) *Through a point within an angle less than 60° there can al-ways be drawn a straight line intersecting both sides of the angle.*
(8) *A circle can be passed through any three noncollinear points.*
(9) *There is no upper limit to the area of a triangle.*

Of the various substitutes, the one most commonly favored in present-day high school geometry texts is (2) above, which was made well known in modern times by the Scottish physicist and mathematician John Playfair (1748–1819), although this particular alternative had been used by others and had even been stated as early as the fifth century by Proclus. It is apparent that many of the substitutes above are hardly any more acceptable than Euclid's postulate, since they are either equally complicated or they assume a geometrical property that is by no means "self-evident."

It constitutes an interesting and challenging set of exercises to try to show the equivalence of the alternatives above to Euclid's original postulate. To show the equivalence of Euclid's postulate and a particular one of the alternatives, one must show that the alternative follows as a theorem from Euclid's assumptions, and also that Euclid's postulate follows as a theorem from Euclid's system of assumptions with the parallel postulate replaced by the considered alternative.

The number of attempts made, throughout the centuries, to derive Euclid's parallel postulate from his other assumptions is almost legion. All these attempts ended in failure, most of them sooner or later being shown to rest upon a tacit assumption equivalent to the postulate itself. The earliest effort of which we are today aware was made by Claudius Ptolemy (ca. A.D. 150). Proclus exposed the fallacy in Ptolemy's attempt by showing that Ptolemy had unwittingly assumed that through a point not on a given straight line only one parallel can be drawn to the line; this assumption is the Playfair equivalent of Euclid's postulate.

Proclus submitted an attempt of his own, but his "proof" rests on the assumption that parallel lines are always a bounded distance apart, and this assumption can be shown to imply Euclid's fifth postulate. Among the more noteworthy attempts of somewhat later times is one made in the thirteenth century by Nasir-ed-din (1201–1272), a Persian astronomer and mathematician who compiled, from an earlier Arabic translation, an improved edition of the *Elements*, and who wrote a treatise on Euclid's postulate, but this attempt, too, involves a tacit assumption equivalent to the postulate being "proved."

An important stimulus to the development of geometry in western Europe after the Renaissance was a renewal of the criticism of Euclid's fifth postulate. Hardly any critical comments are to be found in the early printed editions of the *Elements* made at the end of the fifteenth century and at the beginning of the sixteenth century. However, after the translation, in 1533, of Proclus' *Commentary on Euclid, Book I*, many men once again embarked upon a critical analysis of the fifth postulate. For example, John Wallis (1616–1703), while lecturing at Oxford University, became interested in the work of Nasir-ed-din, and in 1663 offered his own "proof" of the parallel postulate, but his attempt involves the equivalent assumption that similar noncongruent triangles exist. So it was with all the many attempts to derive Euclid's postulate as a theorem; each attempt involved the vitiating circularity of assuming something equivalent to the thing being established or else committed some other form of fallacious reasoning. Most of this vast amount of work is of little real importance in the actual evolution of mathematical thought until we come to the remarkable investigation of the parallel postulate made by Girolamo Saccheri in 1733.

Saccheri was born in San Remo in 1667, showed marked precocity as a youngster, completed his novitiate for the Jesuit Order at the age of twenty-three, and then spent the rest of his life filling a succession of university teaching posts. While instructing rhetoric, philosophy, and theology at a Jesuit college in Milan, Saccheri read Euclid's *Elements* and became enamored with the powerful method of *reductio ad absurdum*. Later, while teaching philosophy at Turin, Saccheri published his *Logica demonstrativa*, in which the chief innovation is the application of the method of *reductio ad absurdum* to the treatment of formal logic. Some years after, while a professor of mathematics at the University of Pavia, it occurred to Saccheri to apply his favorite method of *reductio ad absurdum* to a study of Euclid's parallel postulate. He was well prepared for the task, having dealt ably in his earlier work on logic with such matters as definitions and postulates. Also, he was acquainted with the work of others regarding the parallel postulate and had succeeded in pointing out the fallacies in the attempts of Nasir-ed-din and Wallis.

Saccheri's effort to establish Euclid's parallel postulate by attempting to institute a *reductio ad absurdum* was apparently the first time anyone had conceived the idea of studying the consequences of a denial of the famous postulate. The result of these researches was a little book entitled *Euclides ab omni naevo vindicatus* (*Euclid Freed of Every Flaw*), which was printed in Milan in 1733, only a few months before the author's death. In this work Saccheri accepts the first twenty-eight propositions of Euclid's *Elements*, which, as we have earlier remarked, do not require the fifth postulate for their proof. With the aid of these theorems he then proceeds to study the *isosceles birectangle*, that is, a quadrilateral *ABDC* in which (see Figure 16) $AC = BD$ and the angles at *A* and *B* are right angles. By drawing the diagonals *AD* and *BC* and using simple congruence theorems (which are found among Euclid's first twenty-eight propositions), Saccheri easily shows that the angles at *C* and *D* are equal to each other. But nothing can be ascertained in regard to the magnitude of these angles. Of course, as a consequence of Euclid's fifth postulate, it follows that these angles are both right angles, but the assumption of this postulate is not to be employed. As a result, the two angles might both be right angles, or both be obtuse angles, or both be acute angles. Here Saccheri maintains an open mind, and names the three possibilities the *hypothesis of the right angle,* the *hypothesis of the obtuse angle,* and the *hypothesis of the acute angle.* The plan of the work is to rule out the last two possibilities by show-

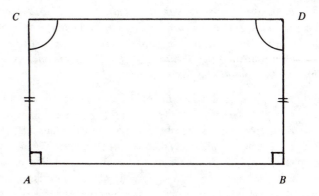

F̲ɪɢ. 16

ing that their respective assumptions lead to contradictions, thus leaving, by *reductio ad absurdum*, the first hypothesis. But this hypothesis can be shown to be equivalent to Euclid's fifth postulate. In this way the parallel postulate is to be established and the blemish of its assumption by Euclid removed.

The task of eliminating the hypothesis of the obtuse angle and the hypothesis of the acute angle turns out to be rather arduous. With real geometrical skill and fine logical penetration, Saccheri establishes a number of theorems, of which the following are among the more important:

1. *If one of the hypotheses is true for a single isosceles birectangular quadrilateral, it is true of every such quadrilateral.*

2. *On the hypothesis of the right angle, the obtuse angle, or the acute angle, the sum of the angles of a triangle is, respectively, equal to, greater than, or less than two right angles.*

3. *If there exists a single triangle for which the sum of the angles is equal to, greater than, or less than two right angles, there follows, respectively, the truth of the hypothesis of the right angle, the obtuse angle, or the acute angle.*

4. *Given a straight line and a point not on the line, on the hypothesis of the right angle, the obtuse angle, or the acute angle, there exists, respectively, precisely one straight line, no straight line, or an infinitude of straight lines through the point which do not meet the given line.*

5. *The locus of the extremity of a perpendicular of constant length which moves with its other end on a fixed straight line is a straight line on the hypothesis of the right angle, a curve convex to the fixed line on the hypothesis of the obtuse angle, and a curve concave to the fixed line on the hypothesis of the acute angle.*

Tacitly assuming, as did Euclid (in the proof of his Proposition I 16), that straight lines are infinite in length, Saccheri managed to eliminate the hypothesis of the obtuse angle, but the case of the hypothesis of the acute angle proved to be much more difficult. After obtaining many of the now classical theorems of so-called non-Euclidean geometry, Saccheri lamely forced into his development an unconvincing contradiction involving hazy notions about elements at infinity. Coming after the careful work that had been presented up to this point, it is difficult to believe that Saccheri himself was really persuaded by his flimsy ending. Had he not been so eager to exhibit a contradiction here, but rather had boldly admitted his inability to find one, Saccheri would today unquestionably be credited with the discovery of non-Euclidean geometry. His work seems to have been little regarded by his contemporaries and was soon forgotten,* and it was not until 1889 that it was dramatically resurrected by his compatriot, Eugenio Beltrami (1835–1900). The principal part of Saccheri's work has been translated into English and can easily be read by any student of elementary geometry.†

In 1766, thirty-three years after Saccheri's publication, Johann Heinrich Lambert (1728–1777) of Switzerland wrote a similar investigation entitled *Die Theorie der Parallellinien*, which, however, was not published until eleven years after his death. Lambert chose as his fundamental figure the *trirectangle*, or quadrilateral containing three right angles, which can be regarded as the half of a Saccheri isosceles birectangle formed by joining the midpoints of the latter's bases. As with Saccheri, three hypotheses arise, according to whether the fourth angle of the trirectangle is right, obtuse, or acute.

Lambert went considerably beyond Saccheri in deducing propositions under the hypotheses of the obtuse and acute angles. Thus, with Saccheri, he showed that in the three hypotheses the sum of the angles of a triangle is equal to, greater than, or less than two right angles, respectively, and then, in addition, that the excess above two right angles in the hypothesis of the obtuse angle, or the deficiency below two right angles in the hypothesis of the acute angle, is proportional to the area of the triangle. He observed the resemblance of the geometry following from the hypothesis of the obtuse angle to

*There is a narrative-rich alternative explanation, involving an insinuation of suppression and confiscation by the Church (brought on by a belated realization of a successfully perpetrated hoax) that has been offered to account for the sudden disappearance and consequent disregard of Saccheri's masterpiece. Did Saccheri all along feel there was no contradiction to be found under the hypothesis of the acute angle, but, with tongue in cheek, put in a ridiculous one that could not possibly fool a mathematician, merely to get his work past the Church censors? See, *e.g.*, E. T. Bell, *The Magic of Numbers*, Chapter 25.

†See G. B. Halsted and D. E. Smith in the list of further readings furnished at the end of this lecture.

spherical geometry, where the area of a triangle is proportional to its spherical excess, and correctly conjectured that the geometry following from the hypothesis of the acute angle can perhaps be verified on a sphere of imaginary radius.

Another notable discovery made by Lambert concerns the measurement of lengths in the two geometries that follow from the obtuse-angle and acute-angle hypotheses. In Euclidean geometry, because similar noncongruent figures exist, lengths can be measured only in terms of some arbitrary unit, such as the foot or the meter, which has no structural connection with the geometry. Angles, on the other hand, possess a natural unit of measure, such as the right angle or the radian, which is capable of geometrical definition. This is what is meant when mathematicians say that in Euclidean geometry lengths are *relative* but angles are *absolute*. Lambert discovered that under the hypotheses of the obtuse and acute angles, angles are still absolute, but lengths also are absolute. In fact, it can be shown for these geometries that for every angle there is a corresponding line segment, so that to a natural unit of measure for angles there corresponds a natural unit of measure for lengths.

Lambert eliminated the hypothesis of the obtuse angle by making the same tacit assumption as had Saccheri, namely, that straight lines are infinite in length, but his conclusions with regard to the hypothesis of the acute angle were indefinite and unsatisfactory. Indeed, it was this incomplete and unsettled state of affairs with regard to the acute hypothesis that held Lambert from publishing his work, with the result that it did not appear until friends put it through the press after his death.

A third distinquished effort to establish Euclid's parallel postulate by the *reductio ad absurdum* method was essayed, over a long period of years, by the eminent French mathematician Adrien-Marie Legendre (1752–1833). He began anew and considered three hypotheses according to whether the sum of the angles of some particular triangle is equal to, greater than, or less than two right angles. Tacitly assuming the infinitude of straight lines, he was able to eliminate the second hypothesis, but, although he made several attempts, he could not dispose of the third hypothesis. His first effort is vitiated by the assumption that the choice of a unit of length will not affect the correctness of his propositions, but this, of course, is equivalent to assuming the existence of similar noncongruent figures. The next attempt is vitiated by assuming the existence of a circle through three noncollinear points. Later Legendre independently observed the fact already discovered by Lambert that, under the third hypothesis, the deficiency of the sum of the angles of a triangle below two right angles is proportional to the area of the triangle. Hence, Legendre reasoned, if by starting with any given triangle one can obtain another triangle containing the given triangle at least twice, then the deficiency for the new triangle will be at least twice the deficiency of the given triangle. By repeating the operation a sufficient number of times, one can finally end with a triangle whose angle sum has become negative, a situation which is absurd. But, in order to solve the problem of obtaining a triangle containing a given triangle twice, Legendre assumed that through a point within a given angle less than 60° there can always be drawn a straight line intersecting both sides of the angle, and this, as we have pointed out earlier, is equivalent to assuming Euclid's fifth postulate.

Legendre's last paper on parallels was published in 1833, the year of his death. He perhaps holds the record for persistence in attempting to prove the famous postulate. Actually, however, Legendre had scarcely progressed as far as had Saccheri a hundred years earlier. Moreover, even before the appearance of his last paper, a Russian mathematician, separated from the rest of the scientific world by barriers of distance and language, had taken a most significant step, the boldness and importance of which were far to transcend anything Legendre had done on the subject.

Before closing our lecture, let us say one more word about Legendre. The bulk of his various endeavors to prove the parallel postulate appeared in successive editions of his very popular *Éléments de géométrie*, which ran from a first edition in 1794 to a twelfth in 1823. This work, which is a pedagogical improvement of Euclid's *Elements* made by considerably rearranging and simplifying the propositions, won high regard in continental Europe and was so favorably received in the United States that it became the prototype of the elementary textbooks in this country. Until quite recent times, every American geometry student learned his geometry from some version of Legendre's work. The first English translation was made in the United States in 1819 by John Farrar of Harvard University. The next English translation was made in 1822 by the famous Scottish litterateur Thomas Carlyle, who early in life was a teacher of mathematics. Carlyle's translation ran through 33 American editions. Since Legendre's efforts to prove the parallel postulate were written in such a simple and straightforward style, and appeared in his widely used *Géométrie*, Legendre managed to create a marked popular interest in the postulate. All in all, Legendre's little geometry book was so influential that, if one were to seek a *pedagogically* important GREAT MOMENT IN MATHEMATICS, the publication of this work would be a prime selection.

Exercises

26.1. Show that Playfair's postulate and Euclid's fifth postulate are equivalent.

26.2. Prove that each of the following statements is equivalent to Playfair's postulate:

(a) If a straight line intersects one of two parallel lines, it will intersect the other also.

(b) Distinct straight lines which are parallel to the same straight line are parallel to one another.

26.3. Replacing Euclid's fifth postulate by the assumption "If two angles of one triangle are equal to two angles of another triangle, then the third angles are also equal," show that the sum of the angles of a triangle is always equal to two right angles.

26.4. Find the fallacy in the following "proof," given by B. F. Thibaut (1809), of Euclid's fifth postulate: Let a straightedge be placed with its edge coinciding with side *CA* of triangle *ABC*. Rotate the straightedge successively about the three vertices *A, B, C,* in the direction *ABC,* so that it coincides in turn with *AB, BC, CA.* When the straightedge returns to its original position it must have rotated through four right angles. But the whole rotation is made up of three rotations equal to the exterior angles of the triangle. It now follows that the sum of the angles of the triangle must be equal to two right angles, and from this follows Euclid's parallel postulate.

26.5. Find the fallacy in the following "proof," given by J. K. F.

Hauff (1819), of Euclid's fifth postulate: Let AD, BE, CF be the altitudes of an equilateral triangle ABC, and let O be the point of concurrency of these altitudes. In right triangle ADC, acute angle CAD equals one half acute angle ACD. Therefore, in right triangle AEO, acute angle OAE equals one half acute angle AOE. A similar statement holds for each of the six small right triangles of which AEO is typical. It now follows that the sum of the angles of triangle ABC is equal to one half the sum of the angles about O, that is, equal to two right angles. But it is known that the existence of a single triangle having the sum of its angles equal to two right angles is enough to guarantee Euclid's fifth postulate.

26.6. Prove, by simple congruence theorems (which do not require the parallel postulate), the following theorems about isosceles birectangles:

(a) The summit angles of an isosceles birectangle are equal to each other.

(b) The line joining the midpoints of the base and summit of an isosceles birectangle is perpendicular to both the base and the summit.

(c) If perpendiculars are drawn from the extremities of the base of a triangle upon the line passing through the midpoints of the two sides, an isosceles birectangle is formed.

(d) The line joining the midpoints of the equal sides of an isosceles birectangle is perpendicular to the line joining the midpoints of the base and summit.

26.7. The hypothesis of the acute angle assumes that the equal summit angles of an isosceles birectangle are acute, or that the fourth angle of a trirectangle is acute. In the following we shall assume the hypothesis of the acute angle:

(a) Let ABC be any right triangle, and let M be the midpoint of the hypotenuse AB. At A construct angle BAD = angle ABC. From M draw MP perpendicular to CB. On AD mark off $AQ = PB$, and draw MQ. Prove triangles AQM and BPM congruent, thus showing that angle AQM is a right angle and that points Q, M, P are collinear. Then $ACPQ$ is a trirectangle with acute angle at A. Now show that, *under the hypothesis of the acute angle, the sum of the angles of any right triangle is less than two right angles.*

(b) Let angle A of triangle ABC be not smaller than either angle B or angle C. Draw the altitude through A, and show, by part (a), that *under the hypothesis of the acute angle, the sum of the angles of any triangle is less than two right angles.*

(c) Consider two triangles, ABC and $A'B'C'$, in which corresponding angles are equal. If $A'B' = AB$, then these triangles are congruent. Suppose $A'B' < AB$. On AB mark off $AD = A'B'$, and on AC mark off $AE = A'C'$. Then triangles ADE and $A'B'C'$ are congruent. Show that E cannot fall on C, since then angle BCA would be greater than angle DEA. Show also that E cannot fall on AC produced, since then DE would cut BC in a point F and the sum of the angles of triangle FCE would exceed two right angles. Therefore E lies between A and C and $BCED$ is a convex quadrilateral. Show that the sum of the angles of this quadrilateral is equal to four right angles. But this is impossible under the hypothesis of the acute angle. It thus follows that we cannot have $A'B' < AB$ and that, *under the hypothesis of the acute angle, two triangles are congruent if the three angles of one are equal to the three angles of the other.* In other words, in the geometry resulting from the acute-angle hypothesis similar figures of different sizes do not exist.

(d) A line segment joining a vertex of a triangle to a point on the opposite side is called a *cevian*. A cevian divides a triangle into two subtriangles, each of which may be similarly divided, and so on. Show that if a triangle is partitioned as above into a finite number of subtriangles, the defect of the original triangle is equal to the sum of the defects of the triangles in the partition.

26.8. Show that a trirectangle can be regarded as half of an isosceles birectangle.

26.9. In one effort to eliminate the hypothesis of the acute angle, Legendre tried to obtain, under this hypothesis, a triangle containing a given triangle at least twice. He proceeded as follows. Let ABC be any triangle such that angle A is not greater than either of the other two angles. Construct on side BC a triangle DCB congruent to triangle ABC, with angle DCB equal to angle B and angle DBC equal to angle C. Through D draw any line cutting AB and AC produced in E and F, respectively. Then triangle AEF contains triangle ABC at least twice.

Show that this construction assumes that through a point within a given angle less than 60° there can always be drawn a straight line intersecting both sides of the angle.

26.10. A *spherical degree* for a given sphere is defined to be any spherical area which is equivalent to (1/720)th of the entire surface area of the sphere. The *spherical excess* of a spherical triangle is defined as the excess, measured in degrees of angle, of the sum of the angles of the triangle above 180°.

(a) Show that the area of a lune whose angle is $n°$ is equal to $2n$ spherical degrees.

(b) Show that the area of a spherical triangle, in spherical degrees, is equal in magnitude to the spherical excess of the triangle.

(c) Show that the area A of a spherical triangle of spherical excess $E°$ is given by

$$A = \frac{\pi r^2 E°}{180°},$$

where r is the radius of the sphere. This shows that, for a given sphere, the area of a spherical triangle is proportional to its spherical excess.

Further Reading

BELL, E. T., *The Magic of Numbers*. New York: McGraw-Hill, 1946.

GANS, DAVID, *An Introduction to Non-Euclidean Geometry*. New York: Academic Press, 1973.

HALSTED, G. B., *Girolamo Saccheri's Euclides Vindicatus*. Chicago: Open Court, 1920.

SMITH, D. E., *A Source Book in Mathematics*. New York; Dover, 1958.

WOLFE, H. E., *Introduction to Non-Euclidean Geometry*. New York: Holt, Rinehart and Winston, 1945.

THE LIBERATION OF GEOMETRY, II

In the previous lecture we saw that, in spite of considerable and prolonged effort, Saccheri, Lambert, and Legendre were unable to find a contradiction under the hypothesis of the acute angle. It is no wonder they found no contradiction under this hypothesis, for it is now known that the geometry developed from a certain basic set of assumptions plus the acute-angle hypothesis is as consistent as the Euclidean geometry developed from the same basic set of assumptions plus the hypothesis of the right angle. In other words, it is now known that the parallel postulate *cannot* be deduced as a theorem from the other assumptions of Euclidean geometry but is *independent* of those other assumptions. It took unusual imagination to entertain such a possibility, for the human mind had for two millennia been bound by the prejudice of tradition to the firm belief that Euclid's geometry was most certainly the only authentic one and that any contrary geometric system simply could not be consistent.

The first to suspect the independence of the parallel postulate were Carl Friedrich Gauss (1777–1855) of Germany, János Bolyai (1802–1860) of Hungary, and Nicolai Ivanovitch Lobachevsky (1793–1856) of Russia. These men independently approached the subject through the Playfair form of the parallel postulate by considering the three possibilities: Through a given point not on a given straight line can be drawn *just one, no,* or *more than one* straight line parallel to the given line. These three situations are equivalent, respectively, to the hypotheses of the right, the obtuse, and the acute angle. Assuming, as did their predecessors, the infinitude of a straight line, the second case was easily eliminated. Inability to find a contradiction in the third case, however, led each of the three mathematicians to suspect, in time, a consistent, though perhaps bizarre, geometry under that hypothesis; and each, unaware of the work of

76

the other two, carried out, for its own intrinsic interest, an extensive development of the new geometry.

Gauss was perhaps the first person really to anticipate a non-Euclidean geometry. Although he meditated a good deal on the matter from very early youth on, probably not until his late twenties did he begin to suspect the parallel postulate to be independent of Euclid's other assumptions. Unfortunately, Gauss failed, throughout his life, to publish anything on the subject, and his advanced conclusions are known to us only through copies of letters to interested friends, a couple of published reviews of works of others, and some notes found among his papers after his death. Although he refrained from publishing his own findings, he strove to encourage others to persist in similar investigations, and it was he who called the new geometry *non-Euclidean.*

Apparently the next person to anticipate a non-Euclidean geometry was János Bolyai, who was a Hungarian officer in the Austrian army and the son of the mathematician Farkas Bolyai, a long-time close personal friend of Gauss. The younger Bolyai undoubtedly received considerable stimulus for his study from his father, who had earlier shown an interest in the problem of the parallel postulate. As early as 1823 János Bolyai began to understand the real nature of the problem that faced him, and a letter written during that year to his father shows the enthusiasm he held for his work. In this letter he discloses a resolution to publish a tract on the theory of parallels as soon as he can find the time and opportunity to put the material in order, and exclaims, "Out of nothing I have created a strange new universe." The father urged that the proposed tract be published with all possible speed, and to assist his son toward that end he offered to include it, under his son's name, as an appendix to a large two-volume semiphilosophical work on elementary mathematics that he was then completing. The expansion and arrangement of ideas proceeded more slowly than János had anticipated, but finally, in 1829, he submitted the finished manuscript to his father, and three years later, in 1832, the tract appeared as a twenty-six-page appendix to the first volume of his father's work.* János Bolyai

---

*For a translation of this appendix, see D. E. Smith, *A Source Book in Mathematics.*

never published anything further on the matter, but he did leave behind a great pile of associated manuscript pages.

Although Gauss and Janós Bolyai are acknowledged to be the first to conceive a non-Euclidean geometry, the Russian mathematician Lobachevsky was actually the first to publish a systematic development of the subject. Lobachevsky spent the greater part of his life at the University of Kazan, first as a student, later as a professor of mathematics, and finally as rector, and his earliest paper on non-Euclidean geometry was published in 1829–30 in the *Kazan Messenger,* two to three years before Bolyai's work appeared in print. This memoir attracted only slight attention in Russia and, because it was written in Russian, practically no attention elsewhere. Lobachevsky followed his initial effort with other publications. For example, in the hope of reaching a wider group of readers, he published, in 1840, a little book written in German entitled *Geometrische Untersuchungen zur Theorie der Parallellinien* (*Geometrical Researches on the Theory of Parallels*),* and then still later, in 1855, a year before his death and after he had become blind, he published in French a final and condensed treatment entitled *Pangéométrie* (*Pangeometry*).† So slowly did information of new discoveries spread in those days that Gauss probably did not hear of Lobachevsky's work until the appearance of the German publication in 1840, and Janós Bolyai was unaware of it until 1848. Lobachevsky himself did not live to see his work accorded any wide recognition, but the non-Euclidean geometry which he developed is nowadays frequently referred to as *Lobachevskian geometry,* and the title of "the Copernicus of geometry" has been conferred upon him.

It was some years after the appearance of the work of Lobachevsky and Bolyai that the mathematical world in general paid much attention to the subject of non-Euclidean geometry, and several decades elapsed before the full implication of the discovery was appreciated. One further thing had first to be achieved, namely, a *proof* of the inner consistency of the new geometry. Although Lobachevsky and Bolyai encountered no contradiction in their extensive investigations

---

*A translation appears in R. Bonola, *Non-Euclidean Geometry: A Critical and Historical Study of Its Development.*

†For a translation see D. E. Smith, *A Source Book in Mathematics.*

of the non-Euclidean geometry based upon the hypothesis of the acute angle, and although they even felt confident that no contradiction would arise, there still remained the possibility that such a contradiction or inconsistency might appear if the investigations should be sufficiently continued. The actual independence of the parallel postulate from the other postulates of Euclidean geometry was not unquestionably established until consistency proofs of the hypothesis of the acute angle were furnished. These were not too long in coming and were supplied by Eugenio Beltrami, Arthur Cayley, Felix Klein, Henri Poincaré, and others. The method was to set up a model of the new geometry within Euclidean geometry, so that the abstract development of the hypothesis of the acute angle would be given a representation in a part of Euclidean space. Then any inconsistency in the non-Euclidean geometry would reflect a corresponding inconsistency in the Euclidean geometry of the representation. The proof is one of *relative* consistency; the Lobachevskian non-Euclidean geometry is shown to be consistent *if* Euclidean geometry is consistent, and, of course, everyone believed Euclidean geometry to be consistent.

One consequence of the consistency of the Lobachevskian non-Euclidean geometry is the final settlement of the ages-old problem of the parallel postulate. The consistency established the fact that the parallel postulate is independent of the other assumptions of Euclidean geometry and proved the impossibility of deducing the postulate as a theorem from those other assumptions. For if the parallel postulate could be deduced, then this very result, being a contradiction of the Lobachevskian parallel postulate, would constitute an inconsistency in the non-Euclidean system.

But there are consequences of the consistency of the non-Euclidean geometry that are much more far-reaching than the settlement of the parallel postulate problem. One of the chief of these is the liberation of geometry from its traditional mold. The postulates of geometry became, for the mathematician, mere hypotheses whose physical truth or falsity need not concern him; the mathematician may take his postulates to suit his pleasure, so long as they be consistent with one another. A postulate, as the word is employed by the mathematician, was seen to have nothing to do with "self-evidence" or "truth." With the possibility of inventing purely "artificial" geometries it became apparent that physical space must be viewed as an empirical concept derived from our external experiences and that the postulates of a geometry designed to describe physical space are simply expressions of this experience, like the laws of a physical science. Euclid's parallel postulate, for example, insofar as it tries to interpret actual space, appears to have the same type of validity as Galileo's law of falling bodies; that is, they are both laws of observation which are capable of verification within the limits of experimental error.

This point of view, that geometry when applied to actual space is an experimental science, or a branch of applied mathematics, is in striking contrast to the Kantian theory of space that dominated philosophical thinking at the time of the discovery of the Lobachevskian geometry. The Kantian theory claimed that space is a framework already existing intuitively in the human mind, that the axioms and postulates of Euclidean geometry are *a priori* judgments imposed on the human mind, and that without these axioms and postulates no consistent reasoning about space can be possible. That this viewpoint is untenable was incontestably demonstrated by the invention of the Lobachevskian geometry.

Indeed, the consistency of the Lobachevskian geometry not only liberated geometry but had a similar effect on mathematics as a whole. Mathematics emerged as an arbitrary creation of the human mind and not as something essentially dictated to us of necessity by the world in which we live. The matter is very neatly put in the following words of E. T. Bell:

> In precisely the same way that a novelist invents characters, dialogues, and situations of which he is both author and master, the mathematician devises at will the postulates upon which he bases his mathematical systems. Both the novelist and the mathematician may be conditioned by their environments in the choice and treatment of their material; but neither is compelled by any extrahuman, eternal necessity to create certain characters or invent certain systems.*

The invention of the non-Euclidean geometry, by puncturing a traditional belief and breaking a centuries-long habit of thought, dealt

---

*E. T. Bell, *The Development of Mathematics,* p. 330.

a severe blow to the *absolute truth* viewpoint of mathematics. In the words of Georg Cantor, "The essence of mathematics lies in its freedom."

Finally, we see that the pattern of material axiomatics, as originated by the Greeks so long ago, now required a considerable revamping to fit the new conception of a branch of mathematics. This important evolution in axiomatics is a GREAT MOMENT IN MATHEMATICS of its own and will be treated in a later lecture.

We have seen that the hypothesis of the obtuse angle was discarded by all who did research in the subject because it contradicted the assumption that a straight line is infinite in length. Recognition of a second non-Euclidean geometry, based on the hypothesis of the obtuse angle, was not achieved until the German mathematician Georg Friedrich Bernhard Riemann (1826–1866) in 1854 discussed the concepts of boundlessness and infiniteness. Although Euclid's Postulate 2 asserts that a straight line may be produced indefinitely, it does not necessarily imply that a straight line is infinite in extent, but merely that it is endless, or boundless. Thus the arc of a great circle joining two points on a sphere may be produced indefinitely along the great circle, making the prolonged arc endless, but certainly it is not infinite in extent. Now it is conceivable that a straight line may behave similarly and that after a finite prolongation it, too, may return upon itself. With Riemann's clarification of the concepts of boundlessness and infiniteness, it was shown that one can realize an internally consistent geometry satisfying the hypothesis of the obtuse angle if Euclid's Postulates 1, 2, and 5 (these were listed in LECTURE 8, and, for convenience, are repeated in the appendix to the present lecture) are modified to read:

1'. Two distinct points determine at least one straight line.
2'. A straight line is boundless.
5'. Any two straight lines in a plane intersect.

This second non-Euclidean geometry has come to be known as *Riemannian non-Euclidean geometry*.

A careful reading of the statements of Euclid's Postulates 1 and 2 will show that they actually say precisely what Postulates 1' and 2' say, though Euclid meant them to imply more. He meant Postulate 1 to imply the *uniqueness* of the straight line determined by the two given points and Postulate 2 to imply the *infiniteness* of a straight line, for he employs these two postulates in these senses, but his statements of the two postulates do not really say that much.

With the liberation of geometry from its bonds of tradition, brought about by the discovery of the Lobachevskian and Riemannian non-Euclidean geometries, the way was open for the invention of a great array of new and interesting geometries, all having postulational bases differing in some manner or other from the postulational foundation of Euclid's geometry. Among these new geometries are *non-Archimedean geometry, non-Desarguesian geometry,* a whole class of *Riemannian geometries* broached by Riemann in 1854, *non-Riemannian geometries, finite geometries* (which contain only a finite number of points, lines, and planes), and many, many more. These new geometries have not proved to be barren of application. For example, Einstein found in his study of the general theory of relativity that he had to adopt a non-Euclidean geometry to describe physical space—it is one of the Riemannian geometries mentioned above. Again, a study conducted in 1947 of *visual space* (the space psychologically observed by persons of normal binocular vision) came to the conclusion that such space can best be described by Lobachevskian non-Euclidean geometry.* Other examples can be given.

Certainly, the liberating discovery of the Lobachevskian and Riemannian non-Euclidean geometries ranks as one of the very greatest of the GREAT MOMENTS IN MATHEMATICS, and we can see why Cassius Keyser proclaimed Euclid's parallel postulate as "perhaps the most famous single utterance in the history of science."†

We closed the preceding lecture with some comments on Legendre's *Éléments de géométrie,* the production of which could well serve as a GREAT MOMENT IN MATHEMATICS in a more extended sequence than we are offering. We close the present lecture with some remarks about another prime candidate for a more protracted sequence of GREAT MOMENTS—it is the work of Riemann referred to above in our lecture.

---

*R. K. Luneberg, *Mathematical Analysis of Binocular Vision.* Princeton, N.J.: Princeton University Press, 1947.

†C. J. Keyser, *Mathematical Philosophy,* p. 113.

Certainly one of the most interesting stories about Riemann concerns his *Habilitationschrift,* or probationary lecture, of 1854—a trial lecture that he had to deliver and have accepted before he could be appointed to an unpaid lectureship at Göttingen University. According to custom, he had to submit titles for three different lectures from which the faculty was to choose one. Expecting that the faculty would follow general practice and pick the first, or possibly the second, of the submitted titles, Riemann assiduously prepared himself for these possibilities, and spent almost no time on the third and rather incautiously submitted title.

Now the third lecture, entitled *Über die Hypothesen welche der Geometrie zu Grunde liegen (On the Hypotheses Which Lie at the Foundations of Geometry),* concerned itself with a topic on which Gauss had pondered for sixty or more years. Gauss was so curious to hear what the brilliant young Riemann might say on this topic that he designated the third lecture as the one for Riemann to deliver. And so, after a frantic eleventh-hour preparation, Riemann presented to the Göttingen faculty, and thus to the world, his third lecture. This lecture proved to be a masterpiece in both mathematics and exposition. It revolutionized the study of geometry and has since been ranked as perhaps the richest single paper of comparable length in the entire history of mathematics. Omitting sticky technical details, so as not to discourage the nonmathematicians of the faculty who were present, this paper of Riemann's cast out an enormous number of new and highly fruitful ideas, ideas that engaged the attention of researchers in subsequent years right up to the present.

In the concluding comments of his lecture, Riemann apologized for presenting such an apparently useless topic, but, he said, the value of such an investigation perhaps lies in its ability to liberate us from preconceived ideas should the time ever come when exploration of physical laws might demand some geometry other than the Euclidean. These highly prophetic words were actually realized some fifty years after his death, through Einstein's general theory of relativity.*

---

*For a translation of Riemann's paper, see D. E. Smith, *A Source Book in Mathematics.*

Appendix

We relist here, for convenience of reference, the five postulates of Euclid's *Elements*.

1. A straight line may be drawn connecting any two given points.
2. A straight line may be produced continuously in a straight line in either direction.
3. A circle may be drawn with any given point as center and passing through any given second point.
4. All right angles are equal to one another.
5. If a straight line falling on two straight lines makes the interior angles on the same side together less than two right angles, the two straight lines, if produced indefinitely, meet on that side on which the angles are together less than two right angles.

Exercises

27.1. Take a fixed circle, Σ, in the Euclidean plane, and interpret the Lobachevskian plane as the interior of Σ, a "point" of the Lobachevskian plane as a Euclidean point within Σ, and a "line" of the Lobachevskian plane as that part of a Euclidean line which is contained within Σ. Verify, in this model, the following statements:

(a) Two distinct "points" determine one and only one "line."

(b) Two distinct "lines" intersect in at most one "point."

(c) Through a "point" P not on a "line" m can be passed infinitely many "lines" not meeting "line" m.

(d) Let the Euclidean line determined by the two "points" P and Q intersect Σ in S and T, in the order S, P, Q, T. Let us interpret the Lobachevskian "distance" from P to Q as

$$\log [(QS)(PT)/(PS)(QT)].$$

If P, Q, R are three "points" on a "line," show that

$$\text{"distance" } PQ + \text{"distance" } QR = \text{"distance" } PR.$$

(e) Let "point" P be fixed and let "point" Q move along a fixed "line" through P toward T. Show that "distance" $PQ \to \infty$.

This model was devised by Felix Klein (1849–1929). With the interpretations above, and with a suitable interpretation of "angle"

between two "lines," it can be shown that all the assumptions necessary for Euclidean geometry, except the parallel postulate, are true
propositions in the geometry of the model. We have seen, in part (c),
that the Euclidean parallel postulate is not such a proposition, but
that the Lobachevskian parallel postulate holds instead. The model
thus proves that the Euclidean parallel postulate cannot be deduced
from the other assumptions of Euclidean geometry, for if it were implied by the other assumptions it would have to be a true proposition
in the geometry of the model.

27.2. Let us interpret the Riemannian non-Euclidean plane as
the surface of a given sphere S, a "point" of the Riemannian plane
as a point on S, and a "line" of the Riemannian plane as a great circle on S. Show that Postulates 1', 2', 5' of the lecture text, along
with Euclid's Postulates 3 and 4, hold in this model, thus establishing the (relative) consistency of the Riemannian non-Euclidean
geometry.

27.3. Consider the following set of postulates about certain objects called "dabbas" and certain collections of dabbas called
"abbas":

P1: Every abba is a collection of dabbas.

P2: There exist at least two dabbas.

P3: If p and q are two dabbas, then there exists one and only one
"abba" containing both p and q.

P4: If L is an abba, then there exists a dabba not in L.

P5: If L is an abba, and p is a dabba not in L, then there exists
one and only one abba containing p and not containing any dabba
that is in L.

(a) Devise a model (or interpretation) of the postulate set to show
that P3 cannot be deduced from the remaining postulates of the set.

(b) Devise a model (or interpretation) of the postulate set to show
that P5 cannot be deduced from the remaining postulates of the set.

(c) Restate the postulates by interpreting "abba" as "straight
line" and "dabba" as "point." Note that P5 is now Playfair's postulate.

27.4. Prove that, in plane Riemannian non-Euclidean geometry,
all the perpendiculars erected on the same side of a given straight

line m are concurrent in a point O, the lengths along these perpendiculars from O to the line m are all equal to one another, and this common length is independent of which straight line in the plane is chosen for m.

27.5. Verify, in spherical geometry, the following theorems of plane Riemannian non-Euclidean geometry:

(a) All perpendiculars erected on the same side of a given straight line m are concurrent in a point O, the lengths of these perpendiculars from O to the line m are all equal to one another, and this common length (call it q) is independent of which straight line in the plane is chosen for m.

(b) If A, B, P are any three points on the line m, then $AP:AB = \angle AOP : \angle AOB$.

(c) All straight lines are finite and of the same constant length $4q$.

(d) Two triangles are congruent if the three angles of one are equal to the three angles of the other.

27.6. Because of the apparently inextricable entanglement of space and matter it may be impossible to determine by astronomical methods whether physical space is Euclidean or non-Euclidean. Since all measurements involve both physical and geometrical assumptions, an observed result can be explained in many different ways by merely making suitable compensatory changes in our assumed qualities of space and matter. For example, it is quite possible that a discrepancy observed in the angle-sum of a triangle could be explained by preserving the assumptions of Euclidean geometry but at the same time modifying some physical law, such as some law of optics. And, again, the absence of any such discrepancy might be compatible with the assumptions of a non-Euclidean geometry, together with some suitable adjustments in our assumptions about matter. On these grounds Henri Poincaré maintained the impropriety of asking which geometry is the true one. To clarify this viewpoint, Poincaré devised an imaginary universe Σ occupying the interior of a sphere of radius R in which he assumed the following physical laws to hold:

(1) At any point P of Σ the absolute temperature T is given by $T = k(R^2 - r^2)$, where r is the distance of P from the center of Σ and k is a constant.

(2) The linear dimensions of a material body vary directly with the absolute temperature of the body's locality.

(3) All material bodies in Σ immediately assume the temperatures of their localities.

(a) Show that it is possible for an inhabitant of Σ to be quite unaware of the three physical laws above holding in his universe.

(b) Show that an inhabitant of Σ would feel that his universe is infinite in extent on the grounds that he would never reach a boundary after taking a finite number N of steps, no matter how large N may be chosen.

(c) Show that geodesics (curves of shortest length joining pairs of points) in Σ are curves bending toward the center of Σ. As a matter of fact, it can be shown that the geodesic through two points A and B of Σ is the arc of a circle or straight line through A and B which cuts the bounding sphere orthogonally.

(d) Let us impose one further physical law on the universe Σ by supposing that light travels along the geodesics of Σ. This condition can be physically realized by filling Σ with a gas having the proper index of refraction at each point of Σ. Show, now, that the geodesics of Σ will "look straight" to an inhabitant of Σ.

(e) Show that in the geometry of geodesics in Σ the Lobachevskian parallel postulate holds, so that an inhabitant of Σ would believe that he lives in a non-Euclidean world. Here we have a piece of ordinary, and supposedly Euclidean, space, which, because of different physical laws, appears to be non-Euclidean.

Further Reading

BELL, E. T., *Men of Mathematics.* New York: Simon & Schuster, 1937.

BONOLA, R., *Non-Euclidean Geometry: A Critical and Historical Study of Its Developments,* trans. by H. S. Carslaw. New York: Dover, 1955.

KEYSER, C. J., *Mathematical Philosophy: A Study of Fate and Freedom.* New York: E. P. Dutton, 1922.

SMITH, D. E., *A Source Book in Mathematics.* New York: Dover, 1958.

WOLFE, H. E., *Introduction to Non-Euclidean Geometry.* New York: Holt, Rinehart and Winston, 1945.

THE LIBERATION OF ALGEBRA, I

At the start of LECTURE 26, we commented that there were two very remarkable and revolutionary mathematical developments that occurred in the first half of the nineteenth century. The first one—the discovery of a non-Euclidean geometry in about 1829—has been discussed in the last two lectures. We now come to the second one—the discovery of a nonconventional algebra in 1843. We shall see that, just as the former development liberated geometry from the traditional geometry of Euclid, the second development liberated algebra from the traditional algebra of the real number system. As in the former case, we shall require two lectures to do the material justice.

While the liberation of geometry found its origins in a critical examination, starting as far back as the days of the ancient Greeks, of Euclid's parallel postulate, the liberation of algebra found its origins in the recognition, first by British mathematicians in the early half of the nineteenth century, of the existence of structure in algebra.

We first clarify what is meant by the phrase *algebraic structure.* In studying the common arithmetic of the positive integers, one encounters two operations, called "addition" and "multiplication." These operations are *binary operations*—that is, to each ordered pair of positive integers *a* and *b* are assigned unique positive integers

---

*The material of LECTURE 28 and LECTURE 29 has been adapted from the fuller treatment given in Chapter 5 of H. Eves and C. V. Newsom, *An Introduction to the Foundations and Fundamental Concepts of Mathematics,* revised edition, Holt, Rinehart and Winston, 1965. This latter material, in turn, was expanded from earlier unpublished notes of the lecture series on GREAT MOMENTS IN MATHEMATICS.

c and d, called, respectively, the *sum* of a and b and the *product* of a and b, and denoted by

$$c = a + b \quad \text{and} \quad d = a \times b.$$

These two binary operations of addition and multiplication performed on the set of positive integers possess certain basic properties. For example, if a, b, c, d denote arbitrary positive integers, we have

1. $a + b = b + a$, the so-called *commutative law of addition*.
2. $a \times b = b \times a$, the *commutative law of multiplication*.
3. $(a + b) + c = a + (b + c)$, the *associative law of addition*.
4. $(a \times b) \times c = a \times (b \times c)$, the *associative law of multiplication*.
5. $a \times (b + c) = (a \times b) + (a \times c)$, the *distributive law of multiplication over addition*.

Now in the early nineteenth century, algebra was regarded simply as symbolized arithmetic. That is, instead of working with specific numbers, as we do in arithmetic, in algebra we employ letters which represent these numbers. Indeed, this is still the view of algebra as taught in the high schools and frequently in the freshman year at college.

The five properties above, then, are statements which always hold in the algebra of positive integers. But, since these statements are symbolic, it is conceivable that they might be applicable to some other set of elements than the positive integers, provided we supply appropriate definitions for the two binary operations involved. This is indeed the case, as the following examples will amply testify. If in each case we denote the set of elements by S, it is an easy matter to verify that the elements, under the two given binary operations of $+$ and $\times$, satisfy all five of these properties. In each example, equality is employed in the sense of identity.

EXAMPLES

(a) Let S be the set of all *even* positive integers, and let $+$ and $\times$ denote the usual addition and multiplication of positive integers.

(b) Let S be the set of all rational numbers (integers and fractions; positive, zero, and negative), and let $+$ and $\times$ denote the usual addition and multiplication of real numbers.

(c) Let S be the set of all real numbers, and let $+$ and $\times$ denote the usual addition and multiplication of real numbers.

(d) Let S be the set of all real numbers of the form $m + n\sqrt{2}$, where m and n are integers, and let $+$ and $\times$ denote the usual addition and multiplication of real numbers.

(e) Let S be the set of *Gaussian integers* (complex numbers $m + ni$, where m and n are ordinary integers and $i = \sqrt{-1}$), and let $+$ and $\times$ denote the usual addition and multiplication of complex numbers.

(f) Let S be the set of all ordered pairs (m, n) of integers, and let $(a, b) + (c, d) = (a + c, b + d)$ and $(a, b) \times (c, d) = (ac, bd)$.

(g) Let S be the set of all ordered pairs (m, n) of integers, and let $(a, b) + (c, d) = (ad + bc, bd)$ and $(a, b) \times (c, d) = (ac, bd)$.

(h) Let S be the set of all ordered pairs (m, n) of integers, and let $(a, b) + (c, d) = (a + c, b + d)$ and $(a, b) \times (c, d) = (ac - bd, ad + bc)$.

(i) Let S be the set of all real polynomials in the real variable x, and let $+$ and $\times$ denote the ordinary addition and multiplication of polynomials.

(j) Let S be the set of all real-valued continuous functions of the variable x defined on the closed interval $0 \le x \le 1$, and let $+$ and $\times$ denote ordinary addition and multiplication of such functions.

(k) Let S be the set consisting of just two distinct elements m and n, where we define

Field
(not ordered)

$$m + m = m, \qquad m \times m = m,$$
$$m + n = n + m = n, \qquad m \times n = n \times m = m,$$
$$n + n = n, \qquad n \times n = n.$$

(l) Let S be the set of all point sets of the plane, and let $a + b$ denote the union of sets a and b, and $a \times b$ the intersection of sets a and b. As a special point set of the plane we introduce an ideal set, the *null set*, which has no points in it.

In view of the above examples, which can easily be extended in number, it is apparent that the five basic properties of the positive integers may also be regarded as properties of many other entirely different systems of elements. The five properties above and their consequences constitute an algebra applicable to the positive integers, but it is evident that the five properties and their consequences also constitute an algebra applicable to many other systems. That is to say, there is a common *algebraic structure* (the five basic properties and their consequences) attached to many different systems. The five basic properties may be regarded as postulates for a particular type of algebraic structure, and any theorem formally implied by these postulates is applicable to each of the examples cited above or to any other interpretation satisfying the five basic properties. Considered from this view, algebra is severed from its tie to arithmetic, and an algebra becomes a purely formal hypothetico-deductive study.

The earliest glimmerings of this modern view of algebra appeared in England, in the first half of the nineteenth century, with the work of George Peacock (1791–1858), a Cambridge graduate and teacher, and later Dean of Ely. Peacock was one of the first to study seriously the fundamental principles of algebra, and in 1830 he published his *Treatise on Algebra*, in which he attempted to give algebra a logical treatment comparable to the treatment of geometry found in Euclid's *Elements*, thus winning for himself the title of "the Euclid of algebra."

Peacock made a distinction between what he called "arithmetic algebra" and "symbolic algebra." The former was regarded by Peacock as the study which results from the use of symbols to denote ordinary *positive decimal numbers*, together with signs for operations, like addition and subtraction, to which these numbers may be subjected. Now, in "arithmetic algebra," certain operations are limited by their applicability. For example, in a subtraction, $a - b$, we must have $a > b$. Peacock's "symbolic algebra," on the other hand, adopts the operations of "arithmetic algebra" but ignores their restrictions. Thus subtraction in "symbolic algebra" differs from the same operation in "arithmetic algebra" in that it is to be regarded as always applicable. The justification of this extension of the rules of "arithmetic algebra" to "symbolic algebra" was called, by Peacock, the *principle of the permanence of equivalent forms*. Peacock's "symbolic algebra" is a universal "arithmetic algebra" whose operations are determined by those of "arithmetic algebra," so far as the two algebras proceed in common, and by the principle of the permanence of equivalent forms in all other cases,

The principle of the permanence of equivalent forms was, in its day, regarded as a powerful concept in mathematics, and it played a historical role in such matters as the early development of the arithmetic of complex numbers and the extension of the laws of exponents from positive integral exponents to exponents of a more general kind. In the theory of exponents, for example, if r is a positive rational number and n is a positive integer, then r^n is, by definition, the product of n factors r. From this definition it readily follows that, for any two positive integers m and n, $r^m r^n = r^{m+n}$. By the principle of the permanence of equivalent forms, Peacock affirmed that in "symbolic algebra," $r^m r^n = r^{m+n}$ no matter what might be the nature of the base r or of the exponents m and n. The hazy principle of the permanence of equivalent forms has today been abandoned, but we are still often guided, when attempting to extend a definition, to formulate the more general definition in such a way that some property or properties of the old definition will be preserved.

British contemporaries of Peacock advance his studies and pushed the notion of algebra closer to the modern concept of the subject. Thus Duncan Farquharson Gregory (1813-1844) published a paper in 1840* in which the commutative and distributive laws in algebra were clearly brought out. Further advances in an understanding of the foundations of algebra were made, in the 1840s, by Augustus De Morgan (1806-1871), another member of the British school of algebraists. In the groping work of the British school can be traced the emergence of the idea of algebraic structure and the preparation for the postulational program in the development of algebra. Soon these ideas of the British school spread to continental Europe, where, in 1867, they were considered with great thoroughness by the German mathematician and historian of mathematics, Herman Hankel (1839-1873). But, even before Hankel's treatment appeared, the Irish mathematician William Rowan Hamilton (1805-1865) and the German mathematician Hermann Günther Grassmann (1809-

---

*D. F. Gregory, "On the real nature of symbolic algebra," *Transactions of the Royal Society of Edinburgh,* 14 (1840), 280.

1877) had published results that were of a considerably more far-reaching character, results that led to the liberation of algebra, in much the same way that the discoveries of Lobachevsky and Bolyai led to the liberation of geometry, and that opened the floodgates of modern abstract algebra. This remarkable work of Hamilton and Grassmann, which can be ranked as a truly GREAT MOMENT IN MATHEMATICS, will be the subject matter of our next lecture.

Before passing to the liberation of algebra by Hamilton and Grassmann, it seems fitting to pause here, out of pedagogical interest, to consider a modern set of postulates for common high school algebra. This structure is today technically known as an *ordered field*.

A *field* is a set S of elements, along with two binary operations on S, here indicated by $\oplus$ and $\otimes$ so that we will not necessarily think of them as ordinary addition and multiplication, satisfying the following postulates.* Equality is used in the sense of identity; thus $a = b$ means a and b are the same element.

P1: *If a and b are in S, then $a \oplus b = b \oplus a$.*

P2: *If a and b are in S, then $a \otimes b = b \otimes a$.*

P3: *If a, b, c are in S, then $(a \oplus b) \oplus c = a \oplus (b \oplus c)$.*

P4: *If a, b, c are in S, then $(a \otimes b) \otimes c = a \otimes (b \otimes c)$.*

P5: *If a, b, c are in S, then $a \otimes (b \oplus c) = (a \otimes b) \oplus (a \otimes c)$ and $(b \oplus c) \otimes a = (b \otimes a) \oplus (c \otimes a)$.*

P6: *S contains an element z* (zero) *such that for any element a of S, $a \oplus z = a$.*

P7: *S contains an element u* (unity), *different from z, such that for any element a of S, $a \otimes u = a$.*

P8: *For each element a in S there exists an element $\bar{a}$ in S such that $a \oplus \bar{a} = z$.*

P9: *If a, b, c are in S, $c \neq z$, and $c \otimes a = c \otimes b$ or $a \otimes c =*

---

*It is customary to place the symbol designating a binary operation between the two elements upon which it operates.

$b \otimes c$, *then* $a = b$. (These are the *cancellation laws* for the operation $\otimes$.)

P10: *For each element* $a \neq z$ *in* S, *there exists an element* a^{-1} *in* S *such that* $a \otimes a^{-1} = u$.

If, in addition to the ten postulates above, the following two postulates hold, the field is called an *ordered field*.

P11: *There exists a subset* P, *not containing* z, *of the set* S *such that if* $a \neq z$ *then one and only one of* a *and* $\overline{a}$ *is in* P.

P12: *If* a *and* b *are in* P, *then* $a \oplus b$ *and* $a \otimes b$ *are in* P.

Definition 1: The elements of P are known as the *positive* elements of S; all other nonzero elements of S are known as the *negative* elements of S.

Definition 2: If a and b are elements of S and if $a \oplus \overline{b}$ is positive, then we write $a \bigodot b$ and $b \bigodot a$.

The postulate set for a field has been made somewhat redundant in order to serve a purpose in our next lecture. For example, in view of Postulate P2, only one of the distributive laws in Postulate P5 needed to be given. Also, it can be shown that the entire Postulate P1 and the entire Postulate P9 are redundant. Note that our original five basic properties of the positive integers appear in Postulates P1, P2, P3, P4, and P5 above.

The twelve postulates for an ordered field constitute a postulate set for elementary algebra. The development of elementary algebra from these postulates would be an edifying, though perhaps tedious, undertaking. Such a development would in all likelihood be too abstract to employ at the high school level, where elementary algebra is first taught, but the idea is tantalizing.

Let us formulate two more definitions.

Definition 3: An element a of S is called an *upper bound* of a nonempty collection M of elements of S if, for each element m of M, either $m \bigodot a$ or $m = a$.

Definition 4: An element a of S is called a *least upper bound* of a nonempty collection M of elements of S if a is an upper bound of M and if $a \bigodot b$ whenever b is any other upper bound of M.

We now define a *complete ordered field* to be an ordered field which further satisfies:

P13: (the postulate of continuity): *If a nonempty collection M of elements of S has an upper bound, then it has a least upper bound.*

It is interesting that the postulates for a complete ordered field constitute a postulate set for the first course in the calculus. In such a treatment, any theorem not utilizing P13 in its proof can be assumed, since such a theorem properly belongs to elementary algebra. Proofs of those theorems requiring P13, however, would be carefully given in full; these are the important theorems of the calculus. A postulational treatment of introductory calculus along these lines has been successfully given at the college freshman level.*

Exercises

28.1. Is the operation $+$ distributive over the operation $\times$ in the set of (a) all integers? (b) Example (l) of the lecture?

28.2. Reduce the left member of each of the following equalities to the right member by using successively an associative, commutative, or distributive law. Following custom, multiplication is here sometimes indicated by a dot ($\cdot$) and sometimes by mere juxtaposition of the factors.

(a) $5(6 + 3) = 3 \cdot 5 + 5 \cdot 6$.
(b) $5(6 \cdot 3) = (3 \cdot 5)6$.
(c) $4 \cdot 6 + 5 \cdot 4 = 4(5 + 6)$.
(d) $a[b + (c + d)] = (ab + ac) + ad$.
(e) $a[b(cd)] = (bc)(ad)$.
(f) $a[b(cd)] = (cd)(ab)$.
(g) $(ad + ca) + ab = a[(b + c) + d]$.
(h) $a + [b + (c + d)] = [(a + b) + c] + d$.

28.3. Actually show that Examples (a) through (l) of the lecture satisfy the five basic properties considered in the lecture.

---

*For such a development see, *e.g.*, E. G. Begle, *Introductory Calculus with Analytic Geometry*, Holt, Rinehart and Winston, 1954. This is a relatively slender calculus text of just over 300 pages.

28.4. Establish the following theorems for any field S:

(a) If $a \oplus z = a$ and $a \oplus z' = a$ for all elements a of S, then $z = z'$. (This proves that the zero element is unique.)

(b) If a, b, c are elements of S and if $a \oplus b = a \oplus c$, then $b = c$. (This is the (left) *cancellation law for the operation* $\oplus$.)

(c) Given two elements a and b of S, then there exists a unique element x of S such that $a \oplus x = b$. (This proves that "subtraction" is always possible in a field.)

(d) If a is any element of S, then $a \otimes z = z$.

(e) If a and b are any two elements of S, then $\bar{a} \otimes \bar{b} = a \otimes b$.

(f) If a and b are elements of S and if $a \otimes b = z$, then either $a = z$ or $b = z$.

(g) Given two elements a and b of S, $a \neq z$, then there exists a unique element x of S such that $a \otimes x = b$. (This proves that "division," except by zero, is always possible in a field.)

(h) Show that $\bar{a} \otimes b = a \otimes \bar{b} = \overline{a \otimes b}$.

28.5. Establish the following theorems for any *ordered* field S:

(a) If a is a positive element of S, then $\bar{a}$ is negative.

(b) Element a of S is positive if and only if $a \ogreaterthan z$.

(c) Element a of S is negative if and only if $a \olessthan z$.

(d) If a and b are distinct elements of S, then either $a \ogreaterthan b$ or $a \olessthan b$, but not both.

(e) If a, b, c are elements of S and if $a \ogreaterthan b$ and $b \ogreaterthan c$, then $a \ogreaterthan c$.

(f) If $a \ogreaterthan b$ and c is any element of S, then $(a + c) \ogreaterthan (b + c)$.

(g) If $a \ogreaterthan b$ and c is positive, then $(a \otimes c) \ogreaterthan (b \otimes c)$.

(h) If a and b are positive elements of S, then (1) $a \otimes \bar{b} = \bar{a} \otimes b =$ the negative element $\overline{a \otimes b}$, (2) $\bar{a} \otimes \bar{b} =$ the positive element $a \otimes b$.

(i) If $a \neq z$ is an element of S, then $a \otimes a$ is positive.

(j) The element u is positive.

28.6. Show that any set of at least two complex numbers, in which the sum, difference, product, and permissible quotient of any two numbers in the set are again in the set, constitutes a field under the binary operations of $+$ and $\times$. (Such a field is known as a *number field*. This explains why ordinary addition, subtraction, multiplication, and division, applied to numbers, are sometimes referred to as the *four field operations.*)

28.7. Deduce Postulate P9 of the lecture text from the other Postulates P2 through P10.

28.8. Which of the Examples (a) through (l) of the lecture are examples of fields?

Further Reading

BEGLE, E. G., *Introductory Calculus with Analytic Geometry.* New York: Holt, Rinehart and Winston, 1954.

PEACOCK, GEORGE, *A Treatise on Algebra.* New York: Scripta Mathematica, 1940.

THE LIBERATION OF ALGEBRA, II

Geometry, as we have seen in LECTURE 27, remained shackled to Euclid's version of the subject until Lobachevsky and Bolyai, in 1829 and 1832, liberated it from its bonds by creating an equally consistent geometry in which one of Euclid's postulates fails to hold. With this accomplishment, a deep-rooted and centuries-old conviction that there can be only the one possible geometry was shattered, and the way was opened for the creation of many new and different geometries.

A similar story can be told of algebra. It seemed inconceivable to the mathematicians of the early nineteenth century that there could exist an algebra different from the common algebra of arithmetic. To attempt, for example, the construction of a self-consistent algebra in which the commutative law of multiplication is denied, not only probably did not occur to anyone of the time, but had it occurred it would have been dismissed as a purely ridiculous idea; after all, how can one possibly have a logical algebra in which $a \times b$ is not equal to $b \times a$? Such was the feeling about algebra when, in 1843, William Rowan Hamilton was forced, by physical considerations, to invent an algebra in which the commutative law of multiplication does not hold. The radical step of abandoning the commutative law did not come easily to Hamilton; it dawned on him only after years of cogitation on a particular problem.

It would carry us too far afield to go into the physical motivation that lay behind Hamilton's creation. Perhaps the best approach, for our purposes, is through Hamilton's elegant treatment of complex numbers as real number pairs, first communicated by him in 1833 to the Royal Irish Academy. To the mathematicians of his time, as indeed is still the case among most college freshman mathematics students of today, a complex number was regarded as a strange hybrid number of the form $a + bi$, where a and b are real numbers and i is some kind of a nonreal number such that $i^2 = -1$, and where addition and multiplication of these numbers are to be accomplished by treating each complex number as a linear polynomial in i, replacing i^2, wherever it might occur, by -1. In this way one finds, for addition,

$$(a + bi) + (c + di) = (a + c) + (b + d)i,$$

and, for multiplication,

$$(a + bi)(c + di) = ac + adi + bci + bdi^2$$
$$= (ac - bd) + (ad + bc)i.$$

If these results should be taken as definitions for addition and multiplication of pairs of complex numbers, it is not difficult to show formally that addition and multiplication are commutative and associative and that multiplication is distributive over addition.

Now, since a complex number $a + bi$ is completely determined by the two real numbers a and b, it occurred to Hamilton to represent the complex number simply and nonmystically by the ordered real number pair (a, b). He defined two such number pairs (a, b) and (c, d) to be equal if and only if $a = c$ and $b = d$. Addition and multiplication of such number pairs he defined (to agree with the above results) to be

$$(a, b) + (c, d) = (a + c, b + d) \quad \text{and}$$
$$(a, b)(c, d) = (ac - bd, ad + bc).$$

With these definitions it is easy to show that addition and multiplication of the ordered real number pairs are commutative and associative and that multiplication is distributive over addition, if one assumes, of course, that these laws hold for the ordinary addition and multiplication of real numbers. In fact, assuming that the real numbers, under ordinary addition and multiplication, satisfy all the postulates of a field (as given in the last lecture), one can show that Hamilton's number pairs, under his definitions of addition and multiplication, also satisfy all the postulates of a field.

It is to be noted that the real number system is *embedded* in the complex number system. By this is meant that if each real number r is identified with the corresponding number pair $(r, 0)$, then the correspondence is preserved under addition and multiplication of complex numbers, for we have

$$(a, 0) + (b, 0) = (a + b, 0) \quad \text{and} \quad (a, 0)(b, 0) = (ab, 0).$$

It follows that, in practice, a complex number of the form $(r, 0)$ can be more simply represented by its corresponding real number r.

To obtain the older form of a complex number from Hamilton's form, we note that any complex number (a, b) can be written as

$$(a, b) = (a, 0) + (0, b) = (a, 0) + (b,0)(0, 1) = a + bi,$$

where $(0, 1)$ is represented by the symbol i, and $(a, 0)$ and $(b, 0)$ are identified with the real numbers a and b. Finally, we see that

$$i^2 = (0, 1)(0, 1) = (-1, 0) = -1.$$

The former mystical aura surrounding complex numbers has been removed, for there is nothing mystical about an ordered pair of real numbers. This was a great achievement on the part of Hamilton.

The complex number system is a very convenient number system for the study of vectors and rotations in the plane.* Hamilton attempted to devise an analogous system of numbers for the study of vectors and rotations in three-dimensional space. In his researches he was led to the consideration, not of ordered real number pairs (a, b) having the real numbers embedded within them, but of ordered real number quadruples (a, b, c, d) having both the real and the complex numbers embedded within them. In other words, defining two such quadruples (a, b, c, d) and (e, f, g, h) to be equal if and only if $a = e$, $b = f$, $c = g$, $d = h$, Hamilton found it necessary to define an addition and multiplication of ordered real number

---

*This convenience results from the fact that, when a complex number $z = a + bi$ is considered as representing the point Z having rectangular Cartesian coordinates (a,b), then the complex number z may also be regarded as representing the vector OZ, where O is the origin of coordinates.

quadruples in such a way that, among other restrictions, he would have

$$(a, 0, 0, 0) + (e, 0, 0, 0) = (a + e, 0, 0, 0),$$
$$(a, 0, 0, 0)(e, 0, 0, 0) = (ae, 0, 0, 0),$$
$$(a, b, 0, 0) + (e, f, 0, 0) = (a + e, b + f, 0, 0),$$
$$(a, b, 0, 0)(e, f, 0, 0) = (ae - bf, af + be, 0, 0).$$

Calling such ordered real number quadruples (real) *quaternions,* Hamilton found that, for his various purposes, he had to formulate the following definitions for addition and multiplication of his quaternions:

$$(a, b, c, d) + (e, f, g, h) = (a + e, b + f, c + g, d + h),$$
$$(a, b, c, d)(e, f, g, h) = (ae - bf - cg - dh,$$
$$af + be + ch - dg,$$
$$ag + ce + df - bh,$$
$$ah + bg + de - cf).$$

It can be shown, with these definitions, that the real numbers and the complex numbers are embedded among the quaternions and that if we identify the quaternion $(m, 0, 0, 0)$ with the real number m, then

$$m(a, b, c, d) = (a, b, c, d)m = (ma, mb, mc, md).$$

It can also be shown that addition of quaternions is commutative and associative, and that multiplication of quaternions is associative and distributive over addition. But the commutative law for multiplication fails to hold. To see this, consider, in particular, the two quaternions $(0, 1, 0, 0)$ and $(0, 0, 1, 0)$. One finds that

$$(0, 1, 0, 0)(0, 0, 1, 0) = (0, 0, 0, 1),$$

while

$$(0, 0, 1, 0)(0, 1, 0, 0) = (0, 0, 0, -1) = -(0, 0, 0, 1);$$

that is, the commutative law for multiplication is broken. In fact, if we represent by the symbols 1, i, j, k, respectively, the *quaternionic units* $(1, 0, 0, 0)$, $(0, 1, 0, 0)$, $(0, 0, 1, 0)$, $(0, 0, 0, 1)$, we can verify that the following multiplication table prevails (that is, the desired product is found in the box common to the <u>row</u> headed by the first factor and the column headed by the second factor).

×	1	i	j	k
1	1	i	j	k
i	i	-1	k	$-j$
j	j	$-k$	-1	i
k	k	j	$-i$	-1

Hamilton has told the story that the idea of abandoning the commutative law of multiplication came to him in a flash, after fifteen years of fruitless meditation, while walking along the Royal Canal near Dublin with his wife just before dusk. He was so struck by the unorthodoxy of the idea that he took out his penknife and scratched the gist of this multiplication table into one of the stones of Brougham Bridge. Today a cement tablet embedded in the stone of the bridge tells the story:

Here as he walked by

on the 16th of October 1843

Sir William Rowan Hamilton

in a flash of genius discovered

the fundamental formula for

quaternion multiplication

$$i^2 = j^2 = k^2 = ijk = -1$$

& cut it in a stone of this bridge

Thus is commemorated for us one of the GREAT MOMENTS IN MATHEMATICS.

We can write the quaternion (a, b, c, d) in the form $a + bi + cj + dk$. When two quaternions are written in this form they may be multiplied like polynomials in i, j, k, and then the resulting product put into the same form by means of the multiplication table above.

In the year 1844, Hermann Günther Grassmann published the first edition of his remarkable *Ausdehnungslehre* (*The Calculus of Extension*), in which he developed classes of algebras of much greater generality than Hamilton's quaternion algebra. Instead of considering just ordered sets of four real numbers, Grassmann considered ordered sets of n real numbers. To each such set $(x_1, x_2, \ldots, x_n)$ Grassmann associated a hypercomplex number of the form $x_1 e_1 + x_2 e_2 + \cdots + x_n e_n$, where $e_1, e_2, \ldots, e_n$ are the fundamental units of his algebra. Two such hypercomplex numbers are added and multiplied like polynomials in $e_1, e_2, \ldots, e_n$. The addition of two such numbers yields, then, a number of the same kind. To make the product of two such numbers a number of the same kind requires the construction of a multiplication table for the units $e_1, \ldots, e_n$ similar to Hamilton's multiplication table for the units $1, i, j, k$. Here one has considerable freedom, and different algebras can be created by making different multiplication tables. The multiplication table is governed by the desired application of the algebra and by the laws of algebra one wishes to preserve.

This is not the place to go deeper into either Hamilton's or Grassmann's work. By developing algebras satisfying laws different from those obeyed by common algebra, these men opened the way for the study of innumerable algebraic structures. By weakening or deleting various postulates of common algebra, or by replacing one or more of the postulates by others that are consistent with the remaining postulates, an enormous variety of systems can be studied. For example,* the deletion of Postulate P2 from the postulates for a field, as listed in the previous lecture, leads to an algebraic structure technically known as a *division ring*, or *sfield*; it can be shown that quaternions constitute a division ring. The deletion of Postulates P2, P7, P10 leads to an *integral domain*; the deletion of Postulates P2,

---

*There are slight variations in these definitions as given by different writers.

P9, P10 to a *ring with unity*; the deletion of Postulates P7, P9, P10 to a *commutative ring*; and the deletion of Postulates P2, P7, P9, P10 to a *ring* (with no qualifying phrase or adjective). In the next lecture, we shall consider a particularly basic and important algebraic structure, obtained in a similar way, known as a *group*. It has been estimated that mathematicians have studied well over 200 such algebraic structures. As the American algebraists Garrett Birkhoff and Saunders Mac Lane have written, "Modern algebra has exposed for the first time the full variety and richness of possible mathematical systems."*

Before closing our lecture, let us consider one more noncommutative algebra—the matric algebra devised by the English mathematician Arthur Cayley (1821-1895) in 1857. Matrices arose with Cayley in connection with linear transformations of the type

$$x' = ax + by,$$
$$y' = cx + dy,$$

where a, b, c, d are real numbers, and which may be thought of as mapping the point (x, y) onto the point (x', y'). Clearly, the transformation above is completely determined by the four coefficients a, b, c, d, and so the transformation can be symbolized by the square array

$$\begin{bmatrix} a & b \\ c & d \end{bmatrix},$$

which we shall call a *(square) matrix (of order* 2). Since two transformations of the kind under consideration are identical if and only if they possess the same coefficients, we define two matrices

$$\begin{bmatrix} a & b \\ c & d \end{bmatrix} \quad \text{and} \quad \begin{bmatrix} e & f \\ g & h \end{bmatrix}$$

to be equal if and only if $a = e$, $b = f$, $c = g$, $d = h$. If the transformation given above is followed by the transformation

---

*Garrett Birkhoff and Saunders Mac Lane, *A Survey of Modern Algebra*, Macmillan, 1941, p. 1.

$$x'' = ex' + fy',$$
$$y'' = gx' + hy',$$

the result can be shown, by elementary algebra, to be the transformation

$$x'' = (ea + fc)x + (eb + fd)y,$$
$$y'' = (ga + hc)x + (gb + hd)y.$$

This motivates the following definition for the product of two matrices:

$$\begin{bmatrix} e & f \\ g & h \end{bmatrix} \begin{bmatrix} a & b \\ c & d \end{bmatrix} = \begin{bmatrix} ea + fc & eb + fd \\ ga + hc & gb + hd \end{bmatrix}.$$

Addition of matrices is defined by

$$\begin{bmatrix} a & b \\ c & d \end{bmatrix} + \begin{bmatrix} e & f \\ g & h \end{bmatrix} = \begin{bmatrix} a + e & b + f \\ c + g & d + h \end{bmatrix},$$

and, if m is any real number, we define

$$m \begin{bmatrix} a & b \\ c & d \end{bmatrix} = \begin{bmatrix} a & b \\ c & d \end{bmatrix} m = \begin{bmatrix} ma & mb \\ mc & md \end{bmatrix}.$$

In the resulting algebra of matrices, it may be shown that addition is both commutative and associative and that multiplication is associative and distributive over addition. But multiplication is not commutative, as is seen by the simple example:

$$\begin{bmatrix} 1 & 0 \\ 0 & 0 \end{bmatrix} \begin{bmatrix} 0 & 1 \\ 0 & 1 \end{bmatrix} = \begin{bmatrix} 0 & 1 \\ 0 & 0 \end{bmatrix},$$

$$\begin{bmatrix} 0 & 1 \\ 0 & 1 \end{bmatrix} \begin{bmatrix} 1 & 0 \\ 0 & 0 \end{bmatrix} = \begin{bmatrix} 0 & 0 \\ 0 & 0 \end{bmatrix}.$$

Although a number of algebras in which multiplication is noncommutative were devised in the middle of the nineteenth century, algebras in which multiplication is nonassociative are, for the most part, of rather recent origin. As examples of such algebras we have *Jordan algebras* and *Lie algebras*. A special Jordan algebra, which is used in quantum mechanics, has matrices for elements, with equality and addition defined as in Cayley's matric algebra, but with the product of two matrices A and B defined as $(AB + BA)/2$, where AB stands for Cayley's product of the two matrices A and B. Although multiplication in this algebra can be shown to be nonassociative, it is obviously commutative. A Lie algebra differs from the above Jordan algebra in that the product of the two matrices A and B is defined by $AB - BA$, where again AB denotes the Cayley product of the matrices A and B. In this algebra, multiplication is neither associative nor commutative.

It is curious that Hamilton's quaternions, which at one time were hailed by many as an indispensable tool for future physicists, have become relegated to little more than a highly interesting museum piece in the history of mathematics; they have become completely supplanted by the considerably more supple vector analysis of the American physicist and mathematician Josiah Willard Gibbs (1839–1903) of Yale University. On the other hand, Cayley's matrices have flourished and today constitute a very important and useful instrument in mathematics. The true fame of quaternions lies in the fact that they broke down the barriers of traditional algebra, rendering their creation an indubitably GREAT MOMENT IN MATHEMATICS.

Exercises

29.1. Determine whether the following binary operations ∗ and | , defined for the positive integers, obey the commutative and associative laws, and whether the operation | is distributive over the operation ∗ .

(a) $a * b = a + 2b$, $a \mid b = 2ab$.
(b) $a * b = a + b^2$, $a \mid b = ab^2$.
(c) $a * b = a^b$, $a \mid b = b$.
(d) $a * b = a^2 + b^2$, $a \mid b = a^2b^2$.

29.2. In Hamilton's treatment of complex numbers as ordered pairs of real numbers, show that:

(a) Addition is commutative and associative.

(b) Multiplication is commutative and associative.

(c) Multiplication is distributive over addition.

(d) $(a, 0) + (b, 0) = (a + b, 0)$.

(e) $(a, 0)(b, 0) = (ab, 0)$.

(f) $(0, b) = (b, 0)(0, 1)$.

(g) $(0, 1)(0, 1) = (-1, 0)$.

29.3. (a) Add the two quaternions $(1, 0, -2, 3)$ and $(1, 1, 2, -2)$.

(b) Multiply, in both orders, the two quaternions $(1, 0, -2, 3)$ and $(1, 1, 2, -2)$.

(c) Show that addition of quaternions is commutative and associative.

(d) Show that multiplication of quaternions is associative and distributive over addition.

(e) Show that the real and complex numbers are embedded within the quaternions.

(f) Multiply the two quaternions $a + bi + cj + dk$ and $e + fi + gj + hk$ like polynomials in i, j, k, and, by means of the multiplication table for the quaternionic units, check into the defined product of the two quaternions.

29.4. (a) If

$$x' = ax + by, \qquad x'' = ex' + fy',$$
$$y' = cx + dy, \qquad y'' = gx' + hy',$$

show that

$$x'' = (ea + fc)x + (eb + fd)y,$$
$$y'' = (ga + hc)x + (gb + hd)y.$$

(b) Given the matrices

$$A = \begin{bmatrix} 2 & -3 \\ 4 & 1 \end{bmatrix}, \qquad B = \begin{bmatrix} -2 & 2 \\ 0 & 3 \end{bmatrix},$$

calculate $A + B$, AB, BA, and A^2.

(c) Show that multiplication of matrices is associative and distributive over addition.

(d) Show that in matric algebra the matrix $\begin{bmatrix} 1 & 0 \\ 0 & 1 \end{bmatrix}$ plays the role of unity, and the matrix $\begin{bmatrix} 0 & 0 \\ 0 & 0 \end{bmatrix}$ plays the role of zero.

(e) Show that

$$\begin{bmatrix} 0 & 1 \\ 0 & 1 \end{bmatrix} \begin{bmatrix} 1 & 0 \\ 0 & 0 \end{bmatrix} = \begin{bmatrix} 0 & 0 \\ 0 & 0 \end{bmatrix}$$

and that

$$\begin{bmatrix} 1 & 0 \\ 0 & 0 \end{bmatrix} \begin{bmatrix} 0 & 1 \\ 0 & 1 \end{bmatrix} = \begin{bmatrix} 1 & 0 \\ 0 & 0 \end{bmatrix} \begin{bmatrix} 0 & 1 \\ 1 & 0 \end{bmatrix}.$$

What two familiar laws of ordinary algebra are broken here?

29.5. (a) Show that the matrix $\begin{bmatrix} 0 & 1 \\ 0 & 0 \end{bmatrix}$ has no square root.

(b) Show that for any real number k,

$$\begin{bmatrix} k & 1+k \\ 1-k & -k \end{bmatrix}^2 = \begin{bmatrix} 1 & 0 \\ 0 & 1 \end{bmatrix},$$

whence the matrix $\begin{bmatrix} 1 & 0 \\ 0 & 1 \end{bmatrix}$ has an infinite number of square roots.

29.6. (a) Show that we may define complex numbers as matrices of the form

$$\begin{bmatrix} a & b \\ -b & a \end{bmatrix},$$

where a and b are real, subject to the usual definitions of addition and multiplication of matrices.

(b) Show that we may define real quaternions as matrices of the form

$$\begin{bmatrix} a + bi & c + di \\ -c + di & a - bi \end{bmatrix},$$

where a, b, c, d are real and $i^2 = -1$, subject to the usual definitions of addition and multiplication of matrices.

29.7. (a) Taking

$$A = \begin{bmatrix} 1 & 0 \\ -1 & 0 \end{bmatrix}, \qquad B = \begin{bmatrix} 1 & 1 \\ -1 & 1 \end{bmatrix}, \qquad C = \begin{bmatrix} 1 & 1 \\ 0 & 1 \end{bmatrix}$$

as elements of a Jordan algebra, calculate $A + B$, AB, BA, $A(BC)$, and $(AB)C$.

(b) Taking A, B, C of part (a) as elements of a Lie algebra, calculate $A + B$, AB, BA, $A(BC)$, and $(AB)C$.

29.8. Consider the set of all ordered real number pairs and define:

    (1) $(a, b) = (c, d)$ if and only if $a = c$ and $b = d$,
    (2) $(a, b) + (c, d) = (a + c, b + d)$,
    (3) $(a, b)(c, d) = (0, ac)$,
    (4) $k(a, b) = (ka, kb)$.

(a) Show that multiplication is commutative, associative, and distributive over addition.

(b) Show that the product of three or more factors is always equal to $(0, 0)$.

(c) Construct a multiplication table for the units $u = (1, 0)$ and $v = (0, 1)$.

Further Reading

BELL, E. T., *Men of Mathematics*. New York: Simon and Schuster, 1937.

CROWE, M. J., *A History of Vector Analysis: The Evolution of the Idea of a Vectorial System*. Notre Dame, Ind.: University of Notre Dame Press, 1967.

EVES, HOWARD, *Elementary Matrix Theory*. Boston: Allyn and Bacon, 1966. Reprinted by Dover, 1979.

FORDER, H. G., *The Calculus of Extension*. New York: Cambridge University Press, 1941.

AN IMPORTANT ATOMIC STRUCTURE

Of the algebraic structures that were developed in the nineteenth century there is one, the so-called *group* structure, that in time came to be recognized as of cardinal importance in mathematics. Though the concept of a group received its first extensive study by Augustin-Louis Cauchy (1789–1857) and his successors, under the particular guise of substitution groups, the concept had been informally utilized as early as 1770 by Joseph Louis Lagrange (1736–1813), and had been given a definition and its name in 1830 by Évariste Galois (1811–1832), in his profound researches in the theory of equations. With the subsequent magnificent work of Cayley, Sylow, Lie, Frobenius (in particular), Klein, Hölder, Poincaré, and others, the study of groups assumed its independent abstract form and developed at a rapid pace. The theory of groups is still, after the three-quarter point of the twentieth century, a very active field of mathematical research.

The creation of the group concept constitutes a GREAT MOMENT IN MATHEMATICS. In the present lecture we will briefly examine the notion of a group and comment on the importance of groups in the field of algebra. In the next lecture we shall consider a stunning application of groups to geometry made by Felix Klein in 1872. This application constitutes another GREAT MOMENT IN MATHEMATICS.

A *group*, which is one of the simplest algebraic structures of consequence, is a nonempty set G of elements in which a binary operation $*$ is defined satisfying the following three postulates:

G1: *For all a, b, c in G, $(a * b) * c = a * (b * c)$.*

G2: *There exists an element i of G such that, for all a in G, $a * i = a$.* (The element i is called a *right identity element* of the group. Later we shall prove that a group possesses only one such element.)

110

G3: *For each element a of G there exists an element a^{-1} of G such that $a * a^{-1} = i$.* (The element a^{-1} is called a *right inverse element* of a. Later we shall prove that an element a of a group possesses only one right inverse element.)

If, in addition to the three postulates above, the following postulate is satisfied, the group is called a *commutative,* or an *Abelian, group.*

G4: *For all a, b in G, $a * b = b * a$.*

A group for which Postulate G4 does *not* hold is called a *non-Abelian group.* If the set G of a group contains only a finite number of distinct elements the group is called a *finite group;* otherwise it is called an *infinite group.* For some purposes the much simpler concept of a *semigroup* is important; it is a nonempty set G of elements in which a binary operation $*$ is defined satisfying the single postulate G1. If, in addition, Postulate G4 is satisfied, the semigroup is called an *Abelian semigroup.* Illustrations of groups are numerous and diverse, as is evident from the following examples.

EXAMPLES

(a) Let G be the set of all integers, and let $*$ denote ordinary addition. Here the integer 0 is the identity element, and the inverse element of a given integer a is its negative. This is an example of an infinite Abelian group.

(b) Let G be the set of all rational numbers with 0 omitted, and let $*$ denote ordinary multiplication. Here the rational number 1 is the identity element, and the inverse element of a given rational number a is its reciprocal $1/a$. This is another example of an infinite Abelian group.

(c) Let G be the set of all translations

$$T: \begin{array}{l} x' = x + h, \\ y' = y + k, \end{array}$$

where h and k are real numbers, of the coordinate plane, and let $T_2 * T_1$ denote the result of performing first translation T_1 and then translation T_2. If T_1 and T_2 are the translations

$$T_1: \quad \begin{matrix} x' = x + h_1, \\ y' = y + k_1, \end{matrix} \qquad T_2: \quad \begin{matrix} x' = x + h_2, \\ y' = y + k_2, \end{matrix}$$

then it is easy to show that $T_2 * T_1$ is the transformation

$$x' = x + (h_1 + h_2),$$
$$y' = y + (k_1 + k_2),$$

which is again a translation. One can easily show that the operation $*$ is associative. The identity element is the translation in which $h = k = 0$, and the inverse of translation T is the translation

$$T^{-1}: \quad \begin{matrix} x' = x - h, \\ y' = y - k. \end{matrix}$$

This, too, is an example of an infinite Abelian group.

(d) Let G be the set of four numbers $1, -1, i, -i$, where $i^2 = -1$, and let $*$ denote ordinary multiplication. Here 1 is the identity element, and the inverse elements of $1, -1, i, -i$ are $1, -1, -i, i$, respectively. This is an example of a finite Abelian group.

(e) Let G be the set of four integers 1, 2, 3, 4, and let $a * b$ denote the remainder obtained by dividing the ordinary product of a and b by 5. We may represent all possible values of $a * b$ by means of the operation table:

*	1	2	3	4
1	1	2	3	4
2	2	4	1	3
3	3	1	4	2
4	4	3	2	1

Here the value of $a * b$ is found in the box common to the row headed by a and the column headed by b. Note that 1 is the identity element, and that the inverse elements of 1, 2, 3, 4 are 1, 3, 2, 4, respectively. This is another example of a finite Abelian group.

An operation table, like the one above, can be made for any finite group. If the table is symmetrical in its principal diagonal, as in the present example, the group is Abelian; otherwise it is non-Abelian. The identity element is the element that heads the column which exactly repeats the column of row headings. To find the inverse of an element a, we merely travel across the row headed by a until we come to the identity element; the element heading the column we are now in is a^{-1}. There is no simple way of checking the associativity of the operation $*$ from the table.

(f) Let G be the set of all rotations of a wheel about its axis through angles which are nonnegative integral multiples of $60°$, where a rotation of $(6n + k)60°$, where n and k are nonnegative integers with $k < 6$, is considered the same as the rotation $k \cdot 60°$. Let $a * b$ denote the rotation b followed by the rotation a. This is an example of a finite Abelian group.

(g) Let G be the set of all 2 by 2 matrices

$$\begin{bmatrix} a & b \\ c & d \end{bmatrix},$$

where a, b, c, d are rational numbers such that $ad - bc \neq 0$, and let $*$ denote ordinary (Cayley) matric multiplication (see LECTURE 29). Here the matrix

$$\begin{bmatrix} 1 & 0 \\ 0 & 1 \end{bmatrix}$$

is the identity element, and the inverse element of matrix

$$\begin{bmatrix} a & b \\ c & d \end{bmatrix}$$

is the matrix

$$\begin{bmatrix} d/(ad - bc) & -b/(ad - bc) \\ -c/(ad - bc) & a/(ad - bc) \end{bmatrix}.$$

This is an example of an infinite non-Abelian group.

(h) Let G be the set of six expressions

$$r, \qquad \frac{1}{r}, \qquad 1 - r, \qquad \frac{1}{1-r}, \qquad \frac{r-1}{r}, \qquad \frac{r}{r-1},$$

and let $a * b$ denote the result of substituting the expression b in place of r in the expression a. For example

$$(1 - r) * \frac{r}{r - 1} = 1 - \frac{r}{r - 1} = \frac{1}{1 - r}.$$

This is an example of a finite non-Abelian group.

A number of things are accomplished by our exhibition of examples of groups. To begin with, the examples establish the existence of both finite and infinite Abelian groups and of both finite and infinite non-Abelian groups. Next, the existence of non-Abelian groups establishes the independence of Postulate G4 from the other three postulates. Finally, our various representations of Abelian groups establish the mutual consistency of Postulates G1, G2, G3, G4. Since some of these representations (see Examples (d), (e), (f)) contain only a finite number of objects, which can be denoted by explicitly displayed symbols, it follows that we have established *absolute* consistency of the postulate set rather than just *relative* consistency. This is the case for any postulate set for which a model can be exhibited containing only a finite number of objects. In contrary situations the method of models can do no more than reduce the consistency of one system to that of another. The important matter of consistency of postulate sets will be considered more fully in a subsequent lecture.

Mathematicians have devised a large number of postulate sets for a group, all of which are, of course, equivalent to one another. Frequently, more is assumed than need be. Thus it is quite common to assume, in G2 and G3, the commutativity of i with every element of G and the commutativity of each element a of G with its inverse element a^{-1}. While, from the point of view of logic, there is no harm in this, it is aesthetically more pleasing to assume as little as possible. We have accordingly adopted the weaker postulate set and shall derive the additional assumptions about commutativity as theorems.* We proceed to do this, leaving some of the proofs as exercises.

---

*Actually, however, our postulate set for a group can itself be further weakened.

Some fundamental theorems of groups

THEOREM 1. *If a, b, c are in G and a $*$ c $=$ b $*$c, then a $=$ b.*

By G3 there exists c^{-1}. From $a * c = b * c$, we then have $(a * c) * c^{-1} = (b * c) * c^{-1}$, or, by G1, $a * (c * c^{-1}) = b * (c * c^{-1})$. Employing G3, we now have $a * i = b * i$, whence finally, by G2, $a = b$.

THEOREM 2. *For all a in G, i $*$ a $=$ a $*$ i.*

By G3 there exists a^{-1}. Hence, applying G1, G3, G2, G3 in turn, we have $(i * a) * a^{-1} = i * (a * a^{-1}) = i * i = i = a * a^{-1}$. By Theorem 1 we then have $i * a = a$. But, by G2, we have $a * i = a$. It now follows that $i * a = a * i$.

THEOREM 3. *A group has a unique identity element.*

Let i and j be two identity elements for the group. Then, by G2 applied to the identity element j, $i * j = i$. Also, by Theorem 2, $i * j = j * i$. But, by G2 applied to the identity element i, $j * i = j$. It now follows that $i = j$.

THEOREM 4. *For each element a of G, $a^{-1} * a = a * a^{-1}$.*

By G1, G3, and Theorem 2, applied in turn, $(a^{-1} * a) * a^{-1} = a^{-1} * (a * a^{-1}) = a^{-1} * i = i * a^{-1}$. Therefore, by Theorem 1, $a^{-1} * a = i$. But, by G3, $a * a^{-1} = i$. It now follows that $a^{-1} * a = a * a^{-1}$.

THEOREM 5. *If a, b, c are in G and c $*$ a $=$ c $*$ b, then a $=$ b.*

THEOREM 6. *Each element of a group has a unique inverse element.*

THEOREM 7. *If a is in G, then $(a^{-1})^{-1} = a$.*

THEOREM 8. *If a and b are in G, then there exist unique elements x and y of G such that a $*$ x $=$ b and y $*$ a $=$ b.*

The importance of groups and semigroups in algebra lies largely in their great cohesive force. Many algebraic systems are actually groups or semigroups with respect to one or more of the binary operations of the systems. In other words, many algebraic structures contain the group structure or the semigroup structure within them as a substructure. Groups and semigroups are like algebraic atoms from which many algebraic systems can be constructed. These ideas are illustrated by the following alternative definitions* of *ring, commutative ring, ring with identity, integral domain, division ring,* and *field* (see LECTURE 29).

— A *ring* is a set S of elements for which two binary operations, here called addition and multiplication, are defined such that (1) S is an Abelian group under addition, with identity element called *zero*, (2) S is a semigroup under multiplication, (3) the two distributive laws of multiplication over addition hold.

— A *commutative ring* is a ring in which the multiplicative semigroup is Abelian.

— A *ring with unity* is a ring in which the multiplicative semigroup has an identity element.

— An *integral domain* is a ring in which the nonzero elements constitute a subsemigroup of the multiplicative semigroup.

— A *division ring* (or *sfield*) is a ring of more than one element in which the nonzero elements constitute a subgroup of the multiplicative semigroup.

— A *field* is a division ring in which the multiplicative semigroup is Abelian. Otherwise stated, a *field* is a set S of at least two elements for which two binary operations, here called addition and multiplication, are defined such that (1) the elements of S constitute an Abelian group under addition, (2) the nonzero elements of S constitute an Abelian group under multiplication, (3) multiplication is both right and left distributive over addition.

Exercises

30.1. Show that the set of two numbers, $1, -1$, under ordinary multiplication, constitutes a group which is a subgroup of that of Example (d) of the lecture.

30.2. (a) Do the even integers form a group with respect to addition?

---

*These definitions may be found in N. Jacobson, *Lectures in Abstract Algebra* (Vol. 1, *Basic Concepts*), Chapter 2.

(b) Do the odd integers form a group with respect to addition?

(c) Do all the rational numbers form a group with respect to multiplication?

(d) Let $a * b = a - b$, where a and b are integers. Do the integers form a group with respect to this operation?

(e) Do all the integral multiples of 3 form a group with respect to addition?

30.3. Let G be the set of all rotations

$$R: \begin{matrix} x' = x \cos \theta - y \sin \theta, \\ y' = x \sin \theta + y \cos \theta, \end{matrix}$$

of the Cartesian plane about the origin, and let $R_2 * R_1$ denote the result of performing first rotation R_1 and then rotation R_2. Show that G, under the operation $*$, constitutes an infinite Abelian group.

30.4. Construct the operation table for Example (d) of the lecture.

30.5. Let G be the set of five integers 0, 1, 2, 3, 4, and let $a * b$ denote the remainder obtained by dividing the ordinary product of a and b by 5. Does G, under the operation $*$, constitute a group?

30.6. Actually show that Example (g) of the lecture constitutes an infinite non-Abelian group.

30.7. (a) Form the operation table for the group of Example (h) of the lecture.

(b) The *cross-ratio* of four collinear points A, B, C, D, considered in this order, is defined to be

$$r = (AB, CD) = \left(\frac{AC}{BC}\right)\left(\frac{BD}{AD}\right).$$

The value of the cross ratio of four collinear points evidently depends upon the order in which we consider the points. There are, then, 24 cross ratios corresponding to the 24 permutations (or orders) of the four points. Show that the number of *distinct* cross ratios is, in general, only six and that they are given by the six expressions of Example (h). For this reason the group of Example (h) is known as the *cross ratio group*.

30.8. Consider the ordered triple (a, b, c). The substitution of c for a, a for b, and b for c, can be represented by the array

$$S = \begin{bmatrix} a & b & c \\ c & a & b \end{bmatrix}.$$

There are, in all, six possible substitutions (counting the identity substitution) that can be made. If S_1 and S_2 denote any two of these six substitutions, let $S_2 * S_1$ denote the result of substitution S_1 followed by substitution S_2. For example

$$\begin{bmatrix} a & b & c \\ b & a & c \end{bmatrix} * \begin{bmatrix} a & b & c \\ a & c & b \end{bmatrix} = \begin{bmatrix} a & b & c \\ b & c & a \end{bmatrix}.$$

Show that the six substitutions of the ordered triple (a, b, c), under the operation $*$, constitute a finite non-Abelian group. This group is known as the *symmetric group of degree 3*; there exists such a substitution group for each positive integral degree n. It was substitution groups that occupied the attention of the early workers in group theory.

30.9. (a) Prove THEOREM 5 of the lecture.
(b) Prove THEOREM 6 of the lecture.
(c) Prove THEOREM 7 of the lecture.
(d) Prove THEOREM 8 of the lecture.

30.10. Show that Postulates G2 and G3 for a group may be replaced by G2′: *If a and b are any elements of G, then there exist elements x and y of G such that $a * x = b$ and $y * a = b$.*

Show that the simpler assumption of just the existence of y such that $y * a = b$ is not sufficient.

30.11. Do we still necessarily have a group if Postulate G3 for a group is replaced by the following? G3′: *For each element a of G there exists an element a^{-1} of G such that $a^{-1} * a = i$.*

Further Reading

JACOBSON, N., *Lectures in Abstract Algebra* (Vol. 1, *Basic Concepts*). Princeton, N.J.: D. Van Nostrand, 1951.

A REMARKABLE CODIFICATION

In 1872, upon appointment to the Philosophical Faculty and the Senate of the University of Erlanger, Felix Klein (1849–1925) delivered, according to custom, an inaugural address in his area of specialty. This address, based upon work by himself and the Norwegian mathematician Sophus Lie (1842–1899) in group theory, set forth a remarkable definition of "a geometry," which served to codify essentially all the existing geometries of the time and pointed the way to new and fruitful avenues of research in geometry. This address, with the program of geometrical study advocated by it, has become known as the *Erlanger Programm*. It appeared right at the time when group theory was invading almost every domain of mathematics and when some mathematicians were beginning to feel that all mathematics is nothing but some aspect or other of group theory.

Klein's application of groups to geometry depends upon the concept of a *nonsingular transformation* of a set S of elements onto itself, by which is simply meant a correspondence under which each element of S corresponds to a unique element of S, and each element of S is the correspondent of a unique element of S. Such a correspondence is said to be *one-to-one*.* If, in a nonsingular transformation T of a set S onto itself, element a of S corresponds to element b of S, we say that under the transformation T, element a *is carried into* element b.

By the *product*, T_2T_1, of two nonsingular transformations T_1 and

---

*More generally, if S and T are two sets, then by a *one-to-one correspondence between the elements of S and the elements of T* we mean a collection of ordered pairs (s,t) such that each element s of S occurs as a first element exactly once, and each element t of T occurs as a second element exactly once.

T_2 of a set S of elements onto itself, we mean the resultant transformation obtained by first performing transformation T_1 and then
transformation T_2. The product of two nonsingular transformations
of a set S onto itself is not necessarily commutative, as is seen by taking T_1 to be a translation of a distance of one unit in the direction of
the positive x-axis applied to the set S of all points in the (x,y)-plane,
and T_2 as a counterclockwise rotation of the set S through $90°$ about
the origin of coordinates. Under T_2T_1 the point $(1,0)$ is carried into
the point $(0,2)$, whereas under T_1T_2 it is carried into the point $(1,1)$,
as is shown in Figure 17. But a product of nonsingular transformations of a set S onto itself is associative, for if T_1, T_2, T_3 are any
three nonsingular transformations of a set S onto itself, $T_3(T_2T_1)$
and $(T_3T_2)T_1$ both denote the resultant transformation of S onto
itself obtained by first performing T_1, then T_2, then T_3. This can be
seen by following the fate under these transformations of some arbitrary element a of S. Thus suppose T_1 carries element a into element b,
T_2 carries element b into element c, and T_3 carries element c into element d. Then T_2T_1 carries element a directly into element c and T_3

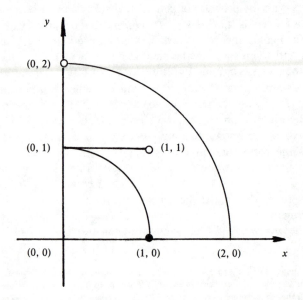

FIG. 17

carries element c into element d, whence $T_3(T_2T_1)$ carries element a directly into element d. On the other hand, T_1 carries element a into element b and T_3T_2 carries element b directly into element d, whence $(T_3T_2)T_1$ also carries element a directly into element d.

Let T be any nonsingular transformation of a set S onto itself, which carries each element a of S into its corresponding element b of S. The transformation which undoes transformation T, by carrying each element b of S back into its original element a of S, is called the *inverse transformation* of transformation T, and is denoted by T^{-1}. The product of T and T^{-1}, taken in either order, is a transformation which clearly leaves all elements of S unchanged; such a transformation is called the *identity transformation*, and will be denoted by I. We note that $TI = T$ for all T.

We may now prove the following important theorem.

THEOREM G. *A nonempty set* Γ *of nonsingular transformations of a set* S *onto itself constitutes a group under multiplication of transformations if* (1) *the product of any two transformations of the set* Γ *is in the set* Γ, (2) *the inverse of any transformation in the set* Γ *is in the set* Γ.

To establish this theorem we first note that, since the product of any two transformations in the set Γ is in the set Γ, multiplication of transformations is a binary operation defined over the set Γ. This binary operation is associative, as we have shown above. Also, if T is a transformation in Γ, then, by property (2), T^{-1} is in Γ. Therefore, by property (1), $I = TT^{-1}$ is in Γ. But, as we have shown above, $TI = T$ and $TT^{-1} = I$ for all T of Γ. All three postulates for a group are thus satisfied, and the theorem is established.

Such a group of nonsingular transformations is called a *transformation group*.

We are now in a position to formulate Felix Klein's famous definition of a geometry, but before doing so it will be interesting and instructive to look into the genesis of the ideas that led Klein to his definition.

In the early part of Euclid's *Elements*, two figures lying in the same unextended plane (that is, in the ordinary plane not augmented by any ideal elements at infinity) are said to be equal if and only if it is possible, by suitable translations, rotations, and reflections in lines, to make one figure coincide with the other. Thus, in the early part of the *Elements*, we study those properties of figures in an unextended plane that are invariant, or unchanged, with respect to the so-called isometries (that is, products of translations, rotations, and reflection in lines) of the plane. But the product of any two planar isometries is a planar isometry, and the inverse of a planar isometry is a planar isometry. It follows, by Theorem G, that the set of all isometries in the plane constitutes a transformation group, and we may consider *plane Euclidean metric geometry* as the study of those properties of figures in an unextended plane that are invariant under the group of planar isometries. Such properties as length, area, congruence, midpoint, parallelism, perpendicularity, collinearity of points, and concurrency of lines are among these invariants, and these are properties studied in plane Euclidean metric geometry.

Plane Euclidean geometry does not concern itself only with properties of figures in an unextended plane that are invariant under planar isometries, for later in the *Element*, similar figures are studied and one becomes concerned with properties of figures that remain invariant under the so-called similarities of the plane. These transformations are compounded from planar isometries and planar homotheties, where, in the latter, each point P is carried into a point P' such that $AP' = k \cdot AP$, A being some fixed point, k some fixed real constant, and A, P, P' collinear. Now the product of two planar similarities is again a planar similarity, and the inverse of a planar similarity is also a planar similarity. It follows, by Theorem G, that the set of all planar similarities constitutes a transformation group, and we may consider *plane similarity geometry*, or *plane equiform geometry*, as the study of those properties of figures of an unextended plane that are invariant under the group of planar similarities. Under this enlarged group, such properties as length, area, and congruence no longer remain invariant and hence are no longer subject to study, but midpoint, parallelism, perpendicularity, collinearity of points, and concurrency of lines are still invariant properties and hence do constitute subject matter for study in this geometry.

Plane Euclidean geometry is thus really a mixture of two more basic geometries, plane Euclidean metric geometry and plane equiform geometry, and each of these is the invariant theory of some associated transformation group.

Plane projective geometry is the study of those properties of figures of an extended plane that remain invariant when the figures are subjected to the so-called projective transformations. Now the product of two projective transformations is again a projective transformation, and the inverse of a projective transformation is also a projective transformation, whence, by Theorem G, the set of all projective transformations constitutes a transformation group. *Plane projective geometry* may be described as the invariant theory of this particular transformation group. Of the previously mentioned properties, only collinearity of points and concurrency of lines still remain invariant. An important invariant of this group of transformations is the cross ratio $(AC/BC)(BD/AD)$ of four collinear points A, B, C, D, and this invariant plays a cardinal role in the study of projective geometry. Since a conic section of any type can be projected into a conic section of any other type, plane projective geometry (in contrast to plane Euclidean geometry) does not study ellipses, parabolas, and hyperbolas as distinct curves.

Theorem G assures us that the set of all projective transformations of an extended plane that carry some fixed line of the plane (usually called the *line at infinity*) onto itself constitutes a transformation group. The study of properties of figures of an extended plane that are invariant under the transformations of this group is known as *plane affine geometry*. Again, Theorem G assures us that the set of all projective transformations of an extended plane that carry a fixed line onto itself and a fixed point not on the line onto itself also constitutes a transformation group. The study of properties of figures of an extended plane that are invariant under the transformations of this group is known as *plane centro-affine geometry*.

The plane non-Euclidean metric geometries, considered in LECTURE 27, can be thought of as the study of those properties of a *non-Euclidean* plane that remain invariant under the group of isometries in those planes.

All the geometries described above are *plane* geometries, but similar studies can be carried out in three-dimensional, or any higher-dimensional, space—each of these higher-dimensional geometries being the invariant theory of some associated transformation group. Again, in all the preceding geometries, the figures upon which the transformations of some transformation group are made to act were considered to be composed of points. Hence, all the preceding geometries are examples of so-called *point* geometries. There are, however, geometries in which entities other than points are chosen as fundamental elements, and geometers have studied many of these, such as line geometry, circle geometry, and sphere geometry. But all these, like the point geometries already considered, can be viewed as the invariant theories of certain transformation groups.

Considerations like these we have been describing led Felix Klein to advance his stunning, highly productive, and very general definition of a geometry, a definition that opened new fields of geometrical research and that introduced a beautiful order into the then-existing chaos of geometrical information. Here is Klein's famous definition of a geometry:

DEFINITION AND NOTATION. A *geometry* is the study of those properties of a set S that remain invariant when the elements of S are subjected to the transformations of some transformation group Γ. Such a geometry will be denoted by the symbol $G(S,\Gamma)$.

In building up a geometry, then, one is at liberty to choose, first of all, the fundamental element of the geometry (point, line, circle, sphere, etc.); next, the set or *space* of these elements (all the points of a plane, all the points of ordinary space, all the points of a spherical surface, all the lines of a plane, all the planes of ordinary space, all the planes of space through a fixed line in space, all the planes of space through a fixed point in space, all the circles of a plane through two fixed points of the plane, the four vertices of a given square, etc.); and, finally, the group of transformations to which the set of elements is to be subjected. The construction of a new geometry becomes, in this way, a rather simple matter. The Erlanger Programm advocated the classification of existing geometries, and the creation and study of new geometries, according to this scheme. In particular, one should study the geometries characterized by the various proper subgroups of the transformation group of a given geometry, in this way obtaining geometries that embrace others. For example, since the transformation group of plane Euclidean metric geometry is a subgroup of the transformation group of plane equiform geometry, it follows that the definitions and theorems holding in the latter geometry must hold in the former. Again, since the transformation group for plane affine geometry is a subgroup of the transformation group for plane projective geometry, it follows that any definition or theorem holding in plane projective geometry also holds in plane affine geometry. By adding the line at infinity to the unextended plane of plane Euclidean metric and plane equiform geometries (this added line will then be considered to transform onto itself in these geometries), it can be seen that certain of the point geometries we have discussed can be arranged in the sequence.

{Euclidean metric, equiform, centro-affine, affine, projective},

wherein the transformation group of any one of the geometries is a proper subgroup of the transformation group of any one of the following geometries in the sequence. Until recent times, the transformation group of projective geometry contained as subgroups the transformation groups of practically all other geometries that had been studied. This is essentially what Arthur Cayley meant when he once remarked that "projective geometry contains all geometry." Actually, so far as the theorems of the geometries are concerned, it is the other way about—the theorems of projective geometry are contained among the theorems of each of the other geometries, and, in the sequence above, the Euclidean metric geometry is the one richest in theorems.

Some geometers prefer to modify slightly Klein's definition of a geometry to read as follows:

MODIFIED DEFINITION. A *geometry* is the study of those properties of a set S that remain invariant when the elements of S are subjected to the transformations of some transformation group Γ, *but which are not invariant under the transformations of a proper supergroup of* Γ.

Under this modified definition, any theorem in a sequence of embracing geometries (like the sequence illustrated above) can belong to only one of the geometries of the sequence, and not to some geometry and also to all those preceding it in the sequence. This permits one to classify a theorem within an embracing sequence of geometries according to the geometry of the sequence to which it belongs.

The various geometries when considered from Klein's point of view can perhaps best be studied by the methods of analytic geometry, where the transformations of each underlying transformation group Γ are given by formulas connecting the new and the old coordinates of the fundamental elements of the space under consideration. For example, the transformation group of plane projective geometry is the totality of all transformations of the form

$$x' = \frac{ax + by + c}{gx + hy + i}, \qquad y' = \frac{dx + ey + f}{gx + hy + i},$$

where a, b, c, ... are any real numbers satisfying the condition

$$\begin{vmatrix} a & b & c \\ d & e & f \\ g & h & i \end{vmatrix} \neq 0.$$

If Γ is taken as the totality of all transformations of the form

$$x' = f(x,y), \quad y' = g(x,y),$$

where $f(x,y)$ and $g(x,y)$ are single-valued and continuous, and where the inverse transformation exists and is also single-valued and continuous, we obtain the rather recently developed branch of geometry known as *topology,* or *analysis situs, of the plane.* From this definition one can see why topology of the plane is frequently referred to as "rubber-sheet" geometry, for in stretching or contracting a rubber sheet the points of the sheet undergo just such a single-valued bicontinuous transformation. Since the product of any two of the transformations above is again a transformation of the same kind, Theorem G assures us that Γ is a transformation group and that topology is a geometry in the sense of Klein's definition.

Topology is perhaps the deepest geometry in the sense considered earlier; its transformation group embraces all previous transformation groups mentioned, and one can amend Cayley's remark to say, "topology contains all geometry." Since the topological transformations are so broad, it is fair to wonder what properties of the plane of points, say, can possibly remain invariant under these transformations. To give a few simple examples, we might mention, first of all, that a simple closed curve (a closed curve that does not cut itself) remains a simple closed curve under all topological transformations. Again, the fact that the deletion of only one point from a simple closed curve does not disconnect the curve is a topological property, and the fact that the deletion of two points from a simple closed curve separates the curve into two pieces is also a topological property. Frequent use is made of topological properties in modern treatments of function theory.

For almost fifty years the Klein synthesis and codification of geometries remained essentially valid. But shortly after the turn of the century, bodies of mathematical propositions, which mathematicians felt should be called geometries, came to light; these bodies of propositions could not be fitted into Klein's codification, and a new point of view upon the matter was developed, based upon the idea of abstract space with a superimposed structure that may or may not be definable in terms of some transformation group. We shall examine this new point of view in a later lecture, merely remarking here that some of these new geometries have found application in the modern theory of physical space that is incorporated in Einstein's general theory of relativity. The Kleinian concept is still highly useful where it applies, and we might call a geometry that fits Klein's definition as given above a *Kleinian geometry*. Partially successful efforts were made in the twentieth century, particularly by the American Oswald Veblen (1880–1960) and the Frenchman Élie Cartan (1869–1951), to extend and generalize Klein's definition so as to include geometries that lie outside Klein's original *Programm*.

Klein's formulation of his fruitful definition of a geometry certainly marks a GREAT MOMENT IN MATHEMATICS. Indeed, so neat and useful is the concept for so many fundamental geometries that the transformation viewpoint has been advocated by a number of pedagogues in connection with the present high school geometry course, and several suitable texts along these lines have appeared.

In summarizing this lecture and the two previous lectures, we can pithily say, with a fair element of truth, that geometry is largely the study of invariants and algebra is largely the study of structure.

Exercises

31.1. Which of the following pairs of nonsingular transformations of the plane of points onto itself are commutative with respect to multiplication?

(a) two translations;

(b) two rotations about the same fixed point O;

(c) a rotation about a fixed point A and another rotation about a different fixed point B;

(d) a rotation about a fixed point A and a reflection in a line m passing through A;

(e) a translation and a reflection in a line m parallel to the direction of the translation;

(f) a translation and a reflection in a line m not parallel to the direction of the translation.

31.2. Prove that the inverse of the product of two nonsingular transformations of a set S onto inself is the product of the inverses of the transformations taken in reverse order; that is, show that $(AB)^{-1} = B^{-1}A^{-1}$. Extend this to the product of any number of nonsingular transformations.

31.3. A nonsingular transformation T of a set S onto itself such that $TT = I$, the identity transformation, is called an *involutoric transformation*.

(a) If T is involutoric, show that $T^{-1} = T$.

(b) Give at least two examples of involutoric transformations of the plane of points onto itself.

31.4. If A and B are two nonsingular transformations of a set S onto itself, then the nonsingular transformation $B' = ABA^{-1}$ is called the *transform of B by A*.

(a) Show that if C' and B' are the transforms of C and B, respectively, by A, then $C'B'$ is the transform of CB by A.

(b) Show that the transform by A of the inverse of B is the inverse of the transform by A of B.

(c) Show that if the product of two nonsingular transformations of a set S onto itself is commutative, then each transformation is its own transform by the other.

(d) Let G_1 be a subgroup of a transformation group G. If every transformation of G_1 is replaced by its transform by T, where T is a fixed transformation of G, show that the transformations thus found form a subgroup of G.

31.5. Show that the set of all nonsingular transformations of the x-axis onto itself of the form

$$ x' = \frac{ax + b}{cx + d}, $$

where a, b, c, d are real and $ad - bc = 1$, constitutes a transformation group.

31.6. (a) Show that the set of all nonsingular transformations of the plane of points onto itself of the form

$$ x' = kx + c, $$
$$ y' = \frac{y}{k} + d, $$

where c and d are real numbers and k is a positive real number, constitutes a transformation group. Such a transformation is called a (special) *planar Lorentz transformation*, and the invariant theory of the group is called a (special) *plane Lorentz geometry*. The Lorentz transformation

$$ x' = kx + c(1 - k), $$
$$ y' = \frac{y}{k} + d\left(1 - \frac{1}{k}\right), $$

is called a *lorotation* (= Lorentz rotation) about the point (c, d); the point (c, d) is called the *center* of the lorotation.

(b) Show that a translation of the plane is a planar Lorentz transformation.

(c) Show that the set of all lorotations about a fixed point (c, d) constitutes a subgroup of the group of all planar Lorentz transformations.

(d) Prove that a lorotation carries a line onto a line, and in particular a line through the center of lorotation onto a line through the center of lorotation.

(e) Prove that a lorotation carries every line onto itself if and only if $k = 1$.

(f) Prove that a lorotation preserves the area of any triangle.

(g) Prove that a lorotation carries a circle with center at the center of lorotation into an ellipse with center at the center of lorotation and with axes parallel to the coordinate axes.

(h) Let us define a *lorcle* (= Lorentz circle) to be the locus of all points obtainable from a given point (a, b) by lorotations about a fixed point (c, d), where $a \neq c$, $b \neq d$. Show that a lorcle is a rectangular hyperbola through the given point whose asymptotes are the parallels to the coordinate axes drawn through the center of lorotation.

31.7. All *direct isometries* of the plane (rotations, translations, and their compounds) constitute a transformation group of the plane of points onto itself. It can be shown that the analytical representation of a general direct isometry is

$$x' = x \cos \theta - y \sin \theta + a,$$
$$y' = x \sin \theta + y \cos \theta + b,$$

where θ, a, b, are arbitrary real numbers. Establish, analytically, the following invariants of the group of direct isometries of the plane:

(a) $(x_1 - x_2)^2 + (y_1 - y_2)^2$, as an invariant of the two points (x_1, y_1), (x_2, y_2). What is the geometrical significance of this invariant?

(b) $(a_1 b_2 - a_2 b_1)/(a_1 b_1 + a_2 b_2)$, as an invariant of the two lines $a_1 x + a_2 y + a_3 = 0$, $b_1 x + b_2 y + b_3 = 0$. What is the geometrical significance of this invariant?

(c) $(a_1 x_0 + a_2 y_0 + a_3)/\sqrt{a_1^2 + a_2^2}$, as an invariant of the point (x_0, y_0) and the line $a_1 x + a_2 y + a_3 = 0$. What is the geometrical significance of this invariant?

(d) $x_0^2 + y_0^2 + a_1 x_0 + a_2 y_0 + a_3$, as an invariant of the point

(x_0, y_0) and the circle $x^2 + y^2 + a_1x + a_2y + a_3 = 0$. What is the geometrical significance of this invariant?

31.8. Consider the following nonsingular transformations of the plane of points onto itself:

R: a clockwise rotation of 90° about the origin,
R':   a clockwise rotation of 180° about the origin,
R'': a clockwise rotation of 270° about the origin,
H: a reflection in the x-axis,
V: a reflection in the y-axis,
D: a reflection in the line $y = x$,
D':   a reflection in the line $y = -x$,
I: the identity motion, in which all points are left unmoved.

(a) Show that the eight transformations constitute a finite non-Abelian transformation group.

(b) Give the inverse of each of the eight transformations.

(c) Imagine a material square having its center at the origin and its sides parallel to the coordinate axes. Show that each of these eight transformations carries this square onto itself. The group of transformations is known as the *group of symmetries of a square*.

Further Reading

GANS, DAVID, *Transformations and Geometries*. New York: Appleton-Century-Crofts, 1969.

ROSSKOPF, M. F., J. L. LEVINE, and B. R. VOGELI, *Geometry, a Perspective View*. New York: McGraw-Hill, 1969.

PYTHAGORAS JUSTIFIED

One fancies that sometime toward the end of the nineteenth century (it is difficult to pinpoint even the precise year) the earth in Western Europe shuddered, and if one had put his ear to the ground he would have heard, rumbling to him in the language of ancient Greece from the far-off grave of the great Pythagoras, "I told you so over two thousand years ago; I told you then that *everything* depends upon the whole numbers." For, after a remarkable sequence of investigations, mathematicians of Western Europe had shown, by the late nineteenth century, that *all* of mathematics is consistent if the natural number system is consistent. Indeed, the great edifice of mathematics was shown to be like an enormous inverted pyramid delicately balanced upon the natural number system as a vertex. The story is a highly interesting and a deeply important one, and the achievement marks a paramount GREAT MOMENT IN MATHEMATICS. For the start of the story we return to the exciting days following the creation of the calculus by Newton and Leibniz some two hundred years earlier.

We have noted, in LECTURE 23, how the wide and amazing applicability of the calculus attracted mathematical researchers of the day and how papers were turned out in great profusion with seemingly little concern regarding the very unsatisfactory foundations of the subject. The operations of the calculus were only poorly understood, and it was natural that, in the feverish application of these operations, paradoxes and absurdities should arise.

Although for almost a century following the creation of the calculus by Newton and Leibniz little was attempted toward logically strengthening the underpinnings of the rapidly growing superstructure of the subject, it must not be supposed there was no criticism of the existing weak base. Lengthy controversies were carried on by some mathematicians, and even the two creators themselves were dissatisfied with their own accounts of the fundamental concepts of the subject. One of the ablest criticisms of the faulty foundations came from a nonmathematician, the eminent metaphysician Bishop George Berkeley (1685–1753), who insisted that the development of the calculus involved the logical fallacy of a shift in the hypothesis. Let us clarify this particular criticism by considering one of Newton's approaches to what is now called differentiation.

In his *Quadrature of Curves* of 1704, Newton determines the derivative (or fluxion, as he called it) of x^3 as follows. We here paraphrase Newton's treatment.

In the same time that x, by growing, becomes $x + o$, the power x^3 becomes $(x + o)^3$, or

$$x^3 + 3x^2o + 3xo^2 + o^3,$$

and the growths, or increments,

$$o \quad \text{and} \quad 3x^2o + 3xo^2 + o^3$$

are to each other as

$$1 \quad \text{to} \quad 3x^2 + 3xo + o^2.$$

Now let the increments vanish, and their last proportion will be 1 to $3x^2$, whence the rate of change of x^3 with respect to x is $3x^2$.

The shift of hypothesis to which Bishop Berkeley objected is evident; in one part of the argument o is assumed to be nonvanishing, while in another part it is taken to be zero. Replies were made to Bishop Berkeley's criticism, but, without a logically rigorous treatment of limits, the objection could not be well met. Alternative approaches proved to be no less confusing. Indeed, all the early explanations of the processes of the calculus are obscure, encumbered with difficulties and objections, and not at all easy to read. Some of the explanations border on the mystical and the absurd, as the statement by Johann Bernoulli that "a quantity which is increased or decreased by an infinitely small quantity is neither increased nor decreased."

Whenever one learns a new mathematical operation, it is imperative also to learn the limitations under which the operation may be performed. Lack of this additional knowledge can lead to the employment of the new operation in a blindly formal manner in situations where the operation is not properly applicable, perhaps resulting in absurd and paradoxical conclusions.

Instructors of mathematics see mistakes of this sort made by their students almost every day. Thus one student of elementary algebra, firmly convinced that $r^0 = 1$ for all real numbers r, will set $0^0 = 1$, while another such student will assume that the equation $ax = b$ always has exactly one real solution for each pair of given real numbers a and b. Again, a student of trigonometry may think that the formula

$$\sqrt{1 - \sin^2 x} = \cos x$$

holds for all real x. A student of the calculus, not aware of improper integrals, may get an incorrect result by apparently correctly applying the rules of formal integration, or he may arrive at a paradoxical result by applying to a certain convergent infinite series some rule which holds only for an *absolutely* convergent infinite series. A good deal of this sort of thing occurred in analysis during the eighteenth century. Attracted by the powerful applicability of the subject, and lacking a real understanding of the foundations upon which the subject must rest, mathematicians manipulated analytical processes in an almost blind manner, often being guided only by a naïve intuition of what was felt must be valid. A gradual accumulation of absurdities was bound to result.

The work of the great Swiss mathematician Leonhard Euler (1707–1783) represents the outstanding example of eighteenth-century formal manipulation in analysis. It was by purely formal devices that Euler discovered the remarkable formula

$$e^{ix} = \cos x + i \sin x,$$

which, for $x = \pi$, yields

$$e^{i\pi} + 1 = 0,$$

a relation connecting five of the most important numbers in mathematics. By formal manipulation Euler arrived at an enormous number of curious relations, like

$$i \log_e i = -\pi/2,$$

and succeeded in showing that any nonzero real number n has an infinite number of logarithms (for a given base), all imaginary if $n < 0$ and all imaginary but one if $n > 0$. The beta and gamma functions of advanced calculus, and many other topics in analysis, similarly originated with Euler. He was a most prolific writer on mathematics and his name is attached to practically every branch of the subject. Although his remarkable mathematical intuition generally held him on the right path, there are instances where his formal manipulation led him into absurdities. For example, if the binomial theorem is formally applied to $(1 - x)^{-1}$, and then x set equal to 2, one finds

$$- 1 = 1 + 2 + 4 + 8 + 16 + \cdots,$$

an absurd result which Euler felt compelled to accept. Also, by adding the two series

$$x + x^2 + \cdots = x/(1 - x)$$

and

$$1 + 1/x + 1/x^2 + \cdots = x/(x - 1),$$

Euler found that

$$\cdots + 1/x^2 + 1/x + 1 + x + x^2 + \cdots = 0.$$

Seventeenth-century and eighteenth-century mathematicians had little understanding of infinite series, and this field of analysis furnished many paradoxes. Thus consider the series

$$S = 1 - 1 + 1 - 1 + 1 - 1 + 1 - 1 + 1 - 1 + \cdots.$$

If we group the terms of this series in one way, we have

$$S = (1 - 1) + (1 - 1) + (1 - 1) + (1 - 1) + \cdots$$
$$= 0 + 0 + 0 + 0 + \cdots = 0,$$

while, if we group the terms in another way, we have

$$S = 1 - (1 - 1) - (1 - 1) - (1 - 1) - (1 - 1) - \cdots$$
$$= 1 - 0 - 0 - 0 - 0 - \cdots = 1.$$

Luigi Guido Grandi (1671-1742) argued that, since the sums 0 and 1 are equally probable, the correct sum of the series is the average value 1/2. This value, too, can be obtained in a purely formal manner, for we have

$$S = 1 - (1 - 1 + 1 - 1 + 1 - 1 + \cdots) = 1 - S,$$

whence $2S = 1$, or $S = 1/2$.

With the gradual accumulation of absurdities and paradoxes within analysis, it became clear that the development of mathematics was undergoing a profoundly disturbing crisis—the second such crisis in the history of the subject. It will be recalled, from LECTURE 5, that the first devastating crisis in the foundations of mathematics occurred in the fifth century B.C., with the unexpected discovery that not all geometrical magnitudes of the same kind are commensurable with one another. This first crisis was finally resolved about 370 B.C. by the brilliant Eudoxus, whose revised theory of magnitude and proportion is one of the great mathematical masterpieces of all time (see LECTURE 6). With the growing realization that the edifice of analysis was being built upon sand, it was inevitable that sooner or later some conscientious mathematicians would feel impelled to attempt the difficult task of establishing a rigorous foundation beneath the calculus. The ultimate accomplishment of this task constitutes, as we shall see, another great achievement in the history of mathematics.

The first really perceptive suggestion of a remedy for the second crisis came from Jean-le-Rond d'Alembert (1717–1783), who, in 1754, remarked that the primitive theory of limits then in use must be replaced by an irreproachably sound theory, but he himself was unable to supply the desired theory. The earliest mathematician of the first rank actually to attempt a rigorization of the calculus was Joseph Louis Lagrange (1736–1813), but his attempt, based upon representing functions by their Taylor's series expansions, was far from successful, for it ignored necessary matters of convergence and divergence. The method was published in 1797 in Lagrange's monumental *Théorie des fonctions analytiques*. Lagrange was one of the leading mathematicians of the eighteenth century, and his work had a deep influence on later mathematical research; with Lagrange's work the long and difficult task of banishing intuition and blind formal manipulation from analysis had begun.

In the nineteenth century, the superstructure of analysis continued to rise, but on ever-deepening foundations. A debt is undoubtedly owed to Carl Friedrich Gauss (1777–1855), for Gauss, more than any other mathematician of his time, broke from intuitive arguments and set new high standards of mathematical rigor. Also, in a treatment of hypergeometric series made by Gauss in 1812, we encounter what is generally regarded as the first really adequate consideration of the convergence of an infinite series.

A great forward stride was made in 1821, when the prolific French mathematician Augustin-Louis Cauchy (1789–1857) successfully met d'Alembert's challenge by developing an acceptable theory of limits and then defining continuity, differentiation, and the definite integral in terms of the limit concept. It is essentially these definitions that we find in the more rigorously written of today's elementary textbooks on the calculus. The limit concept certainly plays a basic role in the development of analysis, for convergence and divergence of infinite series also utilize this concept. Cauchy's rigor inspired other mathematicians to join the effort to rid analysis of formal manipulation and intuitive reasoning.

The demand for an even deeper understanding of the foundations of analysis was strikingly brought out in 1874, when there was exhibited an example, earlier originated by the German mathematician Karl Weierstrass (1815–1897), of a continuous function having no derivative, or, what is the same thing, a continuous curve possessing no tangent at any of its points—something quite contrary to intuitive belief. This example dealt a severe blow to the employment of geometric intuition in analytical studies.

Cauchy's theory of limits, upon which the ideas of continuity, differentiability, and convergence depend, had been built upon a simple intuitive geometrical notion of the real number system. Indeed, Cauchy had taken the real number system more or less for granted, as is still the case in most of our modern elementary calculus texts. It became apparent that the limit concept, continuity, differentiability, and convergence depend upon more recondite properties of the real number system than had been supposed. Other ties connecting analysis with deeper properties of the real number system became evident when Riemann found that Cauchy had unnecessarily restricted his definition of a definite integral; Riemann showed that definite integrals as limits of sums exist even when the integrands

are discontinuous. In elementary calculus these integrals are referred to as *improper integrals,* and they require some care in their
handling. Riemann also produced a function which is continuous for
all irrational values of the variable but discontinuous for all rational
values. Examples such as these made it increasingly apparent that
Cauchy had not struck the true bottom of the difficulties in the way
of a sound foundation for analysis; beneath everything still lay profound properties of the real number system that required urgent understanding.

Accordingly, Weierstrass advocated a program wherein the real
number system itself should first be logically developed, and then
the limit concept, continuity, differentiability, convergence and
divergence, and integrability defined in terms of this number system. This remarkable two-part program became known as the
arithmetization of analysis (so christened by Felix Klein in 1895),
and, though it proved to be intricate and difficult, it was ultimately
realized, toward the end of the nineteenth century, by Weierstrass
and other concerned mathematicians.

The success of Weierstrass's program was far-reaching. In the
first place, since all of analysis can be derived from the real number
system, it follows that all of analysis is consistent if the real number
system is consistent. Now Euclidean geometry, through its Cartesian
interpretation as outlined in LECTURE 20, can also be made to rest
upon the real number system, so it, too, is consistent if the real
number system is consistent; and mathematicians have shown that
most other branches of geometry are consistent if Euclidean geometry is consistent. Again, since the real number system, or some part
of it, can serve for interpreting so many branches of algebra, the
consistency of a good deal of algebra can also be made to depend
upon that of the real number system. It follows that the great bulk of
existing mathematics is consistent if the real number system is consistent. Herein lies the tremendous significance of the real number
system for the foundations of mathematics, and it is no wonder that
Weierstrass's program assumed great importance and that a dedicated group of mathematicians vigorously applied themselves toward its fulfillment.

The second part of Weierstrass's program was accomplished by introducing the meticulous "epsilon-delta" procedures so commonly employed in analysis today. Thus, as an illustration, for the limit concept we have the stark and incisive definition:

If, given any $\epsilon > 0$, there exists a $\delta > 0$ such that $|f(x) - L|$ $< \epsilon$ when $0 < |x - c| < \delta$, then we say L is the limit of $f(x)$ as $x \to c$.

Gone are all such vague phrases as "successive values," "ultimate ratios," "taken as small as one wishes," and "approaches indefinitely close to." Gone also are all references to growing magnitudes and moving points and the abandonment of infinitely small quantities of higher order. All that remains in this precise and unambiguous language and symbolism are real numbers, the operation of addition (and its inverse, subtraction), and the relationship "less than" (and its inverse, "greater than"). In a similar way, all the basic concepts of analysis were carefully formulated in terms of just the real numbers and their fundamental operations and relationships.

Real genius, contributed by a fine array of brilliant mathematicians, was displayed in carrying out the first part of Weierstrass's program. Here were two avenues of approach. One was to obtain a postulate set that characterizes the real numbers. It turned out that the real number system is uniquely determined* as a complete ordered field, and essentially the postulate set listed toward the end of LECTURE 28 was achieved. In that lecture we commented on the pedagogical success of deriving the calculus from this postulate set.

The second approach to a logical development of the real number system was somewhat more subtle than the first approach, and is sometimes called the *genetical,* or *definitional,* approach. Realizing that from the postulates for the real number system we can, by simply making appropriate definitions in terms of the real numbers, arrive at an interpretation of Euclidean geometry, thus making the

---

*That is, in the language of modern abstract algebra, "uniquely determined, *apart from isomorphisms.*"

consistency of Euclidean geometry depend upon that of the real number system, it was wondered if perhaps a similar undertaking can be carried out relative to the real number system itself. That is, perhaps we can start from some postulate set more basic than that for the real number system and, by merely making appropriate definitions, arrive at an interpretation of the real number system. This would make the consistency of the real number system, and hence that of the great bulk of mathematics, depend upon the consistency of the more fundamental postulate set. This was actually accomplished—the real number system was derived purely by adroit definitions, without making any further assumptions, from a postulate set for the much less involved and more basic system of natural numbers, the numbers of simple counting.

The success of carrying out this development is credited to late nineteenth-century researches of Weierstrass, Richard Dedekind (1831–1916), Georg Cantor (1845–1918), Giuseppe Peano (1858–1932), and others. Its accomplishment has given the mathematician a considerable feeling of security concerning the consistency of most of mathematics. This attitude follows from the fact that the natural number system has an intuitive simplicity lacking in most other mathematical systems, and the natural numbers have been very extensively used and studied over a long period of time without uncovering any known inner contradictions.

This is not the place to go into the details of obtaining, without the need of any further postulates, the real number system from the basic system of natural numbers. Much as Hamilton obtained the complex numbers from the real numbers by considering a complex number as an ordered pair of real numbers, so can one obtain the integers from the natural numbers, and then the rational numbers from the integers. To introduce next the irrational numbers in terms of the rational numbers, and thus obtain the completed real number system, is the most difficult part of this approach, but this was accomplished in a number of clever ways.* The step-by-step development can be represented by the scheme

---

*E.g., by Dedekind cuts, by sets of nested intervals with rational endpoints, by convergent sequences of rational numbers, or by decimal representations.

$$N \rightarrow I \rightarrow Q \rightarrow R \rightarrow C,$$

where N represents the system of natural numbers, I the system of integers, Q the system of rational numbers, R the system of real numbers, and C the system of complex numbers. Various other developments are possible, of which a commonly followed program is

$$N \rightarrow Q(+) \rightarrow R(+) \rightarrow R \rightarrow C,$$

where $Q(+)$ represents the system of positive rational numbers and $R(+)$ the system of positive real numbers.

It is interesting that the second step-by-step development indicated above reflects the historical growth of our number system. For mankind first employed the whole, or natural, numbers for counting, then the positive fractions for purposes of measurement; next he introduced positive real numbers to cover incommensurable situations (like the side and diagonal of a square), then acknowledged negative numbers, and, finally, accepted imaginary numbers. On the other hand, the first step-by-step development indicated above reflects the successive algebraic need for new numbers. Thus, if a and b are natural numbers, the equation $x + a = b$ will always have a solution only if we extend the number system to include all integers. Again, if $a \neq 0$ and b are integers, the equation $ax = b$ will always have a solution only if we extend the number system to include all rational numbers. These numbers, though, are not sufficient to yield a solution to the equation $x^2 = 2$, and so the completed system of real numbers is introduced. Finally, the equation $x^2 = -2$ has no solution until the complex numbers are introduced. But here we reach an end, for it has been shown that any polynomial equation having coefficients in the complex number field has a solution within this field—this is the famous so-called *fundamental theorem of* (classical) *algebra*.

Certainly every serious student of mathematics should, sometime or other, work through the details of obtaining the real number system from the natural number system. Upon this small base rests the consistency of the great edifice of mathematics. The ancient Pythagorean belief (at least so far as mathematics is concerned) that everything depends upon the whole numbers is justified, and we see meaning in Leopold Kronecker's often quoted remark, "Die ganzen Zahlen hat Gott gemacht, alles andere ist Menschenwerk."* An historic and truly remarkable GREAT MOMENT IN MATHEMATICS had arrived.

Exercises

32.1. (a) Under what conditions on the real number x is

$$\sqrt{1 - \sin^2 x} = \cos x?$$

(b) Under what conditions does the equation $ax = b$, where a and b are real numbers, have exactly one real solution? no real solution? infinitely many real solutions?

32.2. (a) Since $\sqrt{a}\,\sqrt{b} = \sqrt{ab}$, we have

$$\sqrt{-1}\,\sqrt{-1} = \sqrt{(-1)(-1)} = \sqrt{1} = 1.$$

But, by definition, $\sqrt{-1}\,\sqrt{-1} = -1$. Hence $-1 = +1$. Explain the paradox.

(b) Consider the following indentity, which holds for all values of x and y:

$$\sqrt{x - y} = i\sqrt{y - x}.$$

Setting $x = a$, $y = b$, where $a \neq b$, we find

$$\sqrt{a - b} = i\sqrt{b - a}.$$

Now setting $x = b$, $y = a$, we find

$$\sqrt{b - a} = i\sqrt{a - b}.$$

Multiplying the last two equations, member by member, we get

$$\sqrt{a - b}\,\sqrt{b - a} = i^2\sqrt{b - a}\,\sqrt{a - b}.$$

Dividing both sides by $\sqrt{a - b}\,\sqrt{b - a}$, we finally get

$$1 = i^2, \quad \text{or} \quad 1 = -1.$$

Explain the paradox.

---

*"God made the whole numbers, all the rest is the work of man."

32.3. Find the sum of the roots of the equation

$$(x + 3)/(x^2 - 1) + (x - 3)/(x^2 - x) + (x + 2)/(x^2 + x) = 0.$$

32.4. Most students of elementary algebra will agree to the following theorem: "If two fractions are equal and have equal numerators, then they also have equal denominators." Now consider the following problem. We wish to solve the equation

$$\frac{x + 5}{x - 7} - 5 = \frac{4x - 40}{13 - x}.$$

Combining the terms on the left side, we find

$$\frac{(x + 5) - 5(x - 7)}{x - 7} = \frac{4x - 40}{13 - x}$$

or

$$\frac{4x - 40}{7 - x} = \frac{4x - 40}{13 - x}.$$

By the theorem above, it follows that $7 - x = 13 - x$, or, upon adding x to both sides, that $7 = 13$. What is wrong?

32.5. Explain the following paradox. Certainly

$$3 > 2.$$

Multiplying both sides by log (1/2), we find

$$3 \log (1/2) > 2 \log (1/2)$$

or

$$\log (1/2)^3 > \log (1/2)^2$$

whence

$$(1/2)^3 > (1/2)^2 \quad \text{or} \quad 1/8 > 1/4.$$

32.6. By standard procedure we find

$$\int_{-1}^{1} \frac{dx}{x^2} = \left[-\frac{1}{x} \right]_{-1}^{1} = -1 - 1 = -2.$$

But the function $y = 1/x^2$ is never negative; hence the "evaluation" above cannot be correct. Explain the paradox.

32.7. Let e denote the eccentricity of the ellipse $x^2/a^2 + y^2/b^2 = 1$. It is well known that the length r of the radius vector drawn from the left-hand focus of the ellipse to any point $P(x, y)$ on the curve is given by $r = a + ex$. Now $dr/dx = e$. Since there are no values of x for which dr/dx vanishes, it follows that r has no maximum or minimum. But the only closed curve for which the radius vector has no maximum or minimum is a circle. It follows that every ellipse is a circle. Explain the paradox.

32.8. Explain the paradoxical results concerning the series S of the lecture.

32.9. Any integer, be it positive, zero, or negative, can be represented as the difference $m - n$ of two natural numbers m and n. If $m > n$, then $m - n$ is a positive integer; if $m = n$, then $m - n$ is the zero integer; if $m < n$, then $m - n$ is a negative integer. This suggests the idea of logically introducing the integers as ordered pairs (m, n) of natural numbers, where by (m, n) we actually have in mind the difference $m - n$. With this interpretation of ordered pairs of natural numbers, how should we define equality, addition, and multiplication of these ordered pairs?

32.10. Any rational number (positive, zero, or negative) can be expressed as the quotient m/n of two integers m and n, with $n \neq 0$; in fact, the word *rational* has its origin in this fact. This suggests the idea of logically introducing the rational numbers as ordered pairs (m, n) of integers, with $n \neq 0$, where we actually have in mind the quotient m/n. With this interpretation of ordered pairs of integers, in which the second integer is nonzero, how should we define equality, addition, and multiplication of these ordered pairs?

Further Reading

COURANT, R., and H. ROBBINS, *What Is Mathematics? An Elementary Approach to Ideas and Methods.* New York: Oxford University Press, 1941.

GOFFMAN, C., *Real Functions.* Boston: Prindle, Weber & Schmidt, 1953.

HENKIN, LEON, W. N. SMITH, V. J. VARINEAU, and M. J. WALSH, *Retracing Elementary Mathematics.* New York: Macmillan, 1962.

LANDAU, E. G. H., *Foundations of Analysis: The Arithmetic of Whole, Rational, Irrational and Complex Numbers,* trs. by F. Steinhardt. New York: Chelsea, 1951.

LEVI, H., *Elements of Algebra.* New York: Chelsea, 1954.

RITT, J. F., *Theory of Functions,* rev. ed. New York: King's Crown Press, 1947.

DIGGING DEEPER

One can hardly have failed to notice the increasing use made in the past lectures of the word "set." This word has become one of the most important and basic terms to be found in mathematics. It has become the core of the so-called "new math" program that has permeated the teaching of beginning mathematics, and every school youngster today encounters the set concept over and over again, almost *ad nauseam*, in his mathematics instruction, from very early grade school up through beginning college. There is scarcely a current school mathematics course or textbook that does not start with a discussion of the set idea and of fundamental set notation. Why all this ado about sets? We shall see that there are two reasons for it.

In the preceding lecture we noted that the real number system can be reached from the natural number system in a purely definitional way and that the consistency of the great bulk of mathematics can thus be made to rest upon the consistency of this very basic number system. It is natural to wonder whether the starting point of the definitional development cannot perhaps be pushed to an even deeper level. Mathematicians have shown that this is indeed possible—that one can start with the theory of sets, the concepts of which are already involved in the postulational development of the natural number system, and obtain, in a purely definitional manner, objects that can play the role of the natural numbers. One should, of course, start from a rigorous postulational development of set theory. This is much too extensive a program for us to carry through in detail here, but, assuming an intuitive introduction to set theory, such as all schoolchildren are subjected to, we will be able to indicate how the natural numbers may be defined in terms of set concepts. With the success of this undertaking, the consistency of the great bulk of mathematics will have been shown to rest upon the consistency of set theory. Herein lies one of the two reasons for the paramount importance of sets in mathematics—set theory is, in a sense, a rock bottom foundation for all mathematics.*

We now briefly describe how the natural numbers may be introduced by means of set concepts. We accept the fundamental ideas and notation of set theory taught in the schools today. Some definitions are in order.

DEFINITION 1. Two sets, A and B, are said to be in *one-to-one correspondence* when we have a pairing of the elements of A with the elements of B such that each element of A corresponds to one and only one element of B and each element of B corresponds to one and only one element of A.

There exists, for example, a one-to-one correspondence between the set of all letters of the alphabet and the set of the first 26 positive integers, for we may make the following pairings:

$$a \quad b \quad c \quad d \quad \cdots \quad x \quad y \quad z$$
$$| \quad | \quad | \quad | \qquad | \quad | \quad |$$
$$1 \quad 2 \quad 3 \quad 4 \quad \cdots \quad 24 \quad 25 \quad 26$$

There are, of course, many other permissible pairings of the elements of these two sets; we may pair the 26 successive letters of the alphabet with the successive numbers in any permutation of the first 26 positive integers. As another illustration, there exists a one-to-one correspondence between the set of all positive integers and the set of all even positive integers, for we may make the following pairings:

$$1 \quad 2 \quad 3 \quad 4 \quad \cdots \quad n \quad \cdots$$
$$| \quad | \quad | \quad | \qquad |$$
$$2 \quad 4 \quad 6 \quad 8 \quad \cdots \quad 2n \quad \cdots$$

Also, there exists a one-to-one correspondence between the set of points of a line segment AB and the set of points of any other line segment $A'B'$, for we may make the pairings suggested in Figure 18.

*Logicians have endeavored to push down still further the starting level of the definitional development above by attempting to derive the theory of sets, or classes, from a foundation in the calculus of propositions. The most famous attempt of this kind to date is the monumental *Principia mathematica* of Whitehead and Russell. More will be said about this remarkable work in a later lecture.

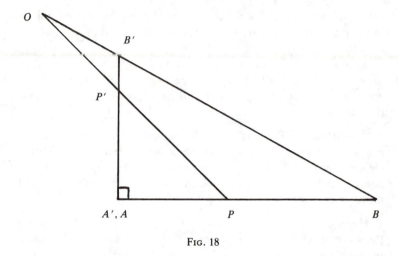

FIG. 18

DEFINITION 2. Two sets, A and B, are said to be *equivalent*, and we write $A \sim B$, if and only if they can be placed in one-to-one correspondence.

Thus the set of all letters of the alphabet and the set of the first 26 positive integers are equivalent, as are the set of all positive integers and the set of all even positive integers, and the set of points on one line segment and the set of points on any other line segment.

We now introduce the concept of number into the theory of sets by the following definition.

DEFINITION 3. Two sets which are equivalent are said to have the *same cardinal number*. All sets that have the same cardinal number as the set $\{a\}$ are said to have *cardinal number one* (or to contain one element); all sets having the same cardinal number as the set $\{a, a'\}$ are said to have the *cardinal number two* (or to contain two elements); all sets having the same cardinal number as the set $\{a, a', a''\}$ are said to have *cardinal number three* (or to contain three elements); and so on. We shall denote the cardinal numbers one, two, three, ... by 1, 2, 3, ...*

---

*This definition presumes ability to recognize distinct objects.

Cardinal numbers thus appear as characteristics of sets. In other words, a triad of apples and a triad of pears have a property in common which we denote by "three." Gottlieb Frege in 1879 and Bertrand Russell in 1901, as have most other logicians who succeeded them, employed the idea of Definition 3 in basic fashion when they defined the cardinal number of a set S as the set of all sets equivalent to set S. Thus, according to this definition, *three* is the set of all sets each of which contains a triad of members.

DEFINITION 4. A nonempty set is said to be *finite* if and only if its cardinal number is one of the cardinal numbers 1, 2, 3, ... A set which is not empty or finite is said to be *infinite*.

It is now possible (and not at all difficult) to show that the finite cardinal numbers as defined above can serve as an interpretation of the natural numbers in any postulate set (of which mathematicians have devised several) that characterizes these natural numbers. It follows, then, that the natural number system is consistent if set theory is consistent, and thence that the great bulk of mathematics is consistent if set theory is consistent. This further deepening of the foundations of mathematics into set theory constitutes another GREAT MOMENT IN MATHEMATICS.

The second reason for the immense importance of set theory in mathematics today is that many mathematical concepts have been considerably generalized and/or succinctly described by set concepts and set notation. We shall give a pair of important instances of this, each instance constituting in itself a GREAT MOMENT IN MATHEMATICS.

We first consider notions of space and the geometry of a space. These notions have undergone marked changes since the days of the ancient Greeks. For the Greeks there was a unique space with a unique geometry; these were absolute concepts. The space was not thought of as a collection of points, but as a realm, or locus, in which objects can be freely moved about and compared with one another. From this point of view, the basic relation in geometry was that of congruence or superimposability.

With the development of analytic geometry in the seventeenth century, space came to be regarded as a collection of points, and with the creation of the classical non-Euclidean geometries in the nineteenth century, mathematicians accepted the situation that there is more than one geometry. But space was still regarded as a locus in which figures can be compared with one another. The central idea became that of a group of congruent transformations of space onto itself, and a geometry came to be regarded as the study of those properties of configurations of points which remain unchanged when the enclosing space is subjected to these transformations. We have seen, in LEC- TURE 31, how this point of view was expanded by Felix Klein in his *Erlanger Programm* of 1872. In the *Erlanger Programm*, a geometry was defined as the invariant theory of a transformation group. This concept synthesized and generalized all earlier concepts of geometry, and supplied a singularly neat classification of a large number of im- portant geometries.

At the end of the nineteenth century there was developed the idea of a branch of mathematics as an abstract body of theorems deduced from a set of postulates, and each geometry became, from this point of view, a particular branch of mathematics. Postulate sets for a large variety of geometries were studied, but the *Erlanger Programm* was in no way upset, for a geometry could be regarded as a branch of mathematics which is the invariant theory of a transformation group.

In 1906, however, Maurice Fréchet (1878–1973) inaugurated the study of abstract spaces, and very general geometries came into being which no longer necessarily fit into the neat Kleinian classification. A space became merely a set of objects, usually called *points*, together with a set of relations in which these points are involved, and a geometry became simply the theory of such a space. The set of rela- tions to which the points are subjected is called the *structure of the space*, and this structure may or may not be explainable in terms of the invariant theory of a transformation group. Although abstract spaces were first formally introduced in 1906, the idea of a geometry as the study of a set of points with some superimposed structure was latent in remarks made by Riemann in his great probationary lecture of 1854.

As an illustration of this further generalization of a space and a geometry of a space, let us describe the abstract spaces created by Fréchet in 1906. In many familiar geometries, such as Euclidean geometry, there is the notion of distance between pairs of points. This idea was generalized by Fréchet, with the aid of set theory concepts, into the study of so-called *metric spaces.*

DEFINITION M. A *metric space* is a set M of elements, called *points,* together with a real number $d(x, y)$, called the *distance function* or *metric* of the space, associated with each pair of points x and y of M, satisfying the following four postulates:

M1. $d(x, y) \geq 0$.
M2. $d(x, y) = 0$ if and only if $x = y$.
M3. $d(x, y) = d(y, x)$.
M4. $d(x, z) \leq d(x, y) + d(y, z)$, where x, y, z are any three, not necessarily distinct, points of M. (This is referred to as the *triangle inequality.*)

Some examples of metric spaces are:
1. The set M of all real numbers in which the distance function is defined by $d(x, y) = |x - y|$. A simple pictorial interpretation of this space is the set of all points on a straight line, with the ordinary concept of distance between two points.
2. The set M of all ordered pairs of real numbers in which the distance function is defined by

$$d(p_1, p_2) = [(x_1 - x_2)^2 + (y_1 - y_2)^2]^{1/2},$$

where

$$p_1 = (x_1, y_1), \qquad p_2 = (x_2, y_2).$$

The student of plane analytic geometry will recognize, as a pictorial interpretation here, the set of all points in a Euclidean plane, together with the Euclidean concept of distance between two points.
3. The set M of all ordered pairs of real numbers in which the distance function is defined by

$$d(p_1, p_2) = |x_2 - x_1| + |y_2 - y_1|,$$

where

$$p_1 = (x_1, y_1), \qquad p_2 = (x_2, y_2).$$

By plotting on a Cartesian plane, one can readily see why this space is frequently referred to as *taxicab space.* Examples 2 and 3 show that quite different metric spaces may possess the same underlying set M.

4. The set M of all infinite sequences $x = \{x_1, x_2, \ldots\}$ of real numbers for which the infinite series $\sum_{i=1}^{\infty} x_i^2$ is convergent, in which the distance function is defined by

$$d(x, y) = \sum_{i=1}^{\infty} [(x_i - y_i)^2]^{1/2}.$$

This example of a metric space is important in a study of real function theory and is known as *Hilbert space.*

We shall not here prove that the four spaces described above are metric spaces. A proof is easy for Examples 1 and 3, but is more difficult for Examples 2 and 4.

Another important abstract space is a so-called *Hausdorff space.*

DEFINITION H. A *Hausdorff space* is a set H of elements, called *points,* together with a collection of certain subsets of these points, called *neighborhoods,* satisfying the following four postulates:

H1. For each point x of H there corresponds at least one neighborhood N_x, where the symbol N_x means that $x \in N_x$.

H2. For any two neighborhoods N_x and N_x' of x, there exists a neighborhood N_x'' such that $N_x'' \subset (N_x \cap N_x')$.

H3. If y is a point of H such that $y \in N_x$, then there is a neighborhood N_y of y such that $N_y \subset N_x$.

H4. If $x \neq y$, there exists an N_x and an N_y such that $N_x \cap N_y = \emptyset$.

To assist in understanding the postulates for a Hausdorff space, let H be the set of all points on a straight line, and select for the neighborhoods of a point x of H the segments of the straight line which have x as midpoint. It is easy to verify that the postulates above are satisfied by this interpretation of points and neighborhoods, and thus we have an example of a Hausdorff space. The arithmetical counterpart of this Hausdorff space is very important in the study of analysis.

If x is any point of a metric space M, we call the subset of all points of M such that $d(x, y) < r$, where r is a positive real number, an *open sphere* with *center* x and *radius* r. It is easy to show that any metric space M can be made into a Hausdorff space H by choosing for neighborhoods all open spheres of M.

As a final example of an abstract space, and one with a very simple and very basic structure, consider a so-called *topological space.*

DEFINITION T. A *topological space** is a set T of elements, called *points,* with a collection of subsets of these points, called *open sets,* satisfying the following three postulates:

T1. T and the null set $\emptyset$ are open sets.
T2. The union of any number of open sets is an open set.
T3. The intersection of any two open sets is an open set.

It can be shown, without difficulty, that a topological space for which there exist, for any two distinct points x and y, disjoint open sets S_x and S_y containing x and y, respectively, is a Hausdorff space with respect to the open sets as neighborhoods.

Abstract spaces have become very important in modern mathematical studies, and their introduction by Fréchet in 1906 certainly marks a GREAT MOMENT IN MATHEMATICS.

Having given some attention to the matter of individual sets possessing a superimposed structure, we now briefly look at pairs of related sets. A considerable amount of mathematical theory is concerned with two related sets. This leads to the concept of *function* and, consequently, to the elaborate branch of mathematics known as *function theory.*

The concept of function, like the notions of space and geometry, has undergone an interesting evolution. Students of mathematics often encounter various refinements of this evolution as their studies progress from the elementary and intuitive courses of high school into the advanced and sophisticated courses of the graduate college level.

The history of the notion of function furnishes an example of the tendency of mathematicians to generalize and extend their concepts. The word "function," in its Latin equivalent, seems to have been introduced by Leibniz (1646–1716) in 1694, at first as a term to denote any quantity connected with a curve, such as the coordinates of a point on the curve, the slope of the curve, the radius of curvature of the curve, and so on. Johann Bernoulli (1667–1748), by 1718, had

---

*The definition of a topological space is not uniform in the literature.

come to regard a function as any expression made up of a variable and some constants, and Euler (1707-1783), somewhat later, regarded a function as any equation or formula involving variables and constants. This latter idea is the notion of a function formed by many students of elementary mathematics courses. The familiar notation $f(x)$ was not used at first, but entered about 1734 with A. C. Clairaut (1713-1765) and Euler.

The Eulerian concept of function remained unchanged until Fourier (1768-1830), in 1807, was led, in his investigations of heat flow, to consider the so-called trigonometric series. These series, as we have seen in LECTURE 25, involve a more general type of relationship between variables than had previously been studied, and, in an attempt to furnish a definition of function broad enough to encompass such relationships, Lejeune Dirichlet (1805-1859), in 1837, perfecting a viewpoint earlier held by Augustin-Louis Cauchy (1789-1857), arrived at the following formulation: A *variable* is a symbol which represents any one of a set of numbers; if two variables x and y are so related that whenever a value is assigned to x there is automatically assigned, by some rule or correspondence, a value to y, then we say y is a (single-valued) *function* of x. The variable x, to which values are assigned at will, is called the *independent variable*, and the variable y, whose values depend upon those of x, is called the *dependent variable*. The permissible values that x may assume constitute the *domain of definition* of the function, and the values taken on by y constitute the *range of values* of the function. It has been customary for the student of mathematics to meet the Dirichlet definition of function in his introductory course in calculus. The definition is a very broad one and does not imply anything regarding the possibility of expressing the relationship between x and y by some kind of analytic expression; it stresses the basic idea of a relationship between two sets of numbers.

Set theory has naturally extended the concept of function to embrace relationships between any two sets of elements, be the elements numbers or anything else. Thus, in set theory, a function f is defined to be any set of ordered pairs of elements such that if $(a_1, b_1) \in f$, $(a_2, b_2) \in f$, and $a_1 = a_2$, then $b_1 = b_2$. The set A of all first elements of the ordered pairs is called the *domain* of the function, and the set B of all second elements of the ordered pairs is called the *range* of the

function. A functional relationship is thus nothing but a special kind
of subset of the Cartesian product set $A \times B$. A one-to-one correspondence is, in turn, a special kind of function, namely, a function
f such that if $(a_1, b_1) \in f$, $(a_2, b_2) \in f$, and $b_1 = b_2$, then $a_1 = a_2$. If,
for a functional relationship f, $(a, b) \in f$, we write $b = f(a)$, which we
read, *"b* equals f at a."

The notion of function pervades much of mathematics, and since
the early part of the present century various influential mathematicians have advocated the employment of this concept as the unifying
and central principle in the organization of elementary mathematics
courses. The concept seems to form a natural and effective guide for
the selection and development of textual material. There is no doubt
of the value of a mathematics student's early acquaintance with the
function concept, and the generalization given to the concept by set
theory marks a GREAT MOMENT IN MATHEMATICS.

The property of continuity may immediately be associated with the
concept of function if the study pertains to spaces A and B defined by
a system of neighborhoods. In fact, a function f may be said to be *continuous* at a point a of A if, for any neighborhood $N_b \subset B$, where b is
the point of B associated with the given point a of A, there exists a
neighborhood $N_a \subset A$ such that, for any $z \in N_a$, $f(z) \in N_b$.

In summarizing the present lecture, we may point out that we have
noted two outstanding accomplishments of the late nineteenth and
early twentieth centuries: (1) set theory may be taken as a foundation
of mathematics, (2) many mathematical concepts can be generalized
and given succinct definitions in terms of set concepts and notation.
Because of these two remarkable accomplishments, many pedagogues
have felt that set theory should assume a prime and basic position in
early mathematical instruction. Thus was created the so-called "new
math" program in our schools. While there is much to say on behalf
of this program, it must be confessed that in the zeal of enthusiasm
many sins were committed in carrying it out. The aftermath of the
new math reveals that we must not drag sets and set notation into
school mathematics just for the sake of doing so but that we should
employ these concepts only when they truly serve to clarify and
simplify the development. In short, let us maintain sanity and judg-
ment here. Is it, for example, better to say, in beginning geometry,
"two coplanar straight lines are said to be parallel if and only if they

do not intersect," or to say "two straight lines m and n such that $m \subset$
π and $n \subset \pi$, where π is some plane, are said to be parallel if and only
if $m \cap n = \varnothing$"?

Exercises

33.1. Whether an endpoint A (or B) of a line segment AB is to be
considered as belonging or not belonging to the segment will be indicated by using a bracket or a parenthesis, respectively, about the
letter A (or B). Using this notation, show that the segments $[AB]$,
$(AB]$, $[AB)$, (AB), considered as sets of points, are equivalent to one
another.

33.2. Show that the set of points comprising a finite segment and
the set comprising an infinite segment are equivalent to one another.

33.3. Dedekind defined an *infinite* set as one that is equivalent to a
proper subset of itself. Show, by Dedekind's definition, that the set of
points comprising a line segment is an infinite set.

33.4. Show that if $d(x, y)$ is a metric for a set M of points, then we
may also use as a metric for M
(a) $kd(x, y)$, where k is a positive real number,
(b) $\sqrt{d(x, y)}$,
(c) $d(x, y)/[1 + d(x, y)]$.

33.5. (a) Show that any set M of elements can be made into a
metric space by setting $d(x, y) = 1$ if $x \neq y$, and $d(x, y) = 0$ if $x = y$.
(b) Show that the set M of all ordered pairs of real numbers, in
which the distance function is defined by

$$d(p_1, p_2) = \max(|x_2 - x_1|, |y_2 - y_1|),$$

where $p_1 = (x_1, y_1)$, $p_2 = (x_2, y_2)$, is a metric space.

33.6. (a) Show that the set of all points in the plane can be made
into a Hausdorff space by selecting for neighborhoods of a point P the
interiors of all circles having P as center.
(b) Show that the set of all points in the plane can be made into a
Hausdorff space by selecting for neighborhoods of a point P the interiors of all squares having centers at P and having sides parallel to
two given perpendicular lines of the plane.

(c) Show that any set of points can be made into a Hausdorff space if we select for neighborhoods the points themselves.

33.7. (a) Let T be the three distinct points A, B, C, and let the open sets of T be $\emptyset$, $\{A\}$, $\{A,B\}$, $\{A,C\}$, T. Show that T is a topological space.

(b) Show that T in part (a) is not a Hausdorff space if the open sets of T are chosen as neighborhoods.

33.8. A point x of a Hausdorff space H is called a *limit point* of a subset S of H if and only if every neighborhood of x contains at least one point of S distinct from x. Prove that any neighborhood N_x of a limit point x of a set S of a Hausdorff space contains an infinite number of points of S.

33.9. Prove that Examples 1 and 3 of the lecture text are metric spaces.

33.10. (a) Show that any subset of a metric space is, for the same metric, a metric space.

(b) Show that any metric space M can be made into a Hausdorff space H by choosing for neighborhoods all open spheres of M.

(c) Show that a topological space for which there exist, for any two distinct points x and y, disjoint open sets S_x and S_y containing x and y, respectively, is a Hausdorff space with respect to the open sets as neighborhoods.

33.11. Show that the four postulates for a metric space may be replaced by just the following two:

M'1. $d(x, y) = 0$ if and only if $x = y$.

M'2. $d(x, z) \le d(z, y) + d(y, x)$, where x, y, z are any three, not necessarily distinct, points of M.

33.12. In elementary mathematics, functions of a limited sort only are considered. Thus a typical definition of "function" in elementary mathematics might read as follows: A *function* is a rule which assigns one and only one number to each number in some collection of numbers.

(a) Show that this definition is a special case of the more general set-theory definition given in the lecture text.

(b) What constitutes the *domain of definition* in this elementary concept of a function?

(c) What constitutes the *range of values* of a function as defined in elementary mathematics?

(d) In what sense can the equation $F = (9/5)C + 32$ be regarded as a function in elementary mathematics?

(e) In what sense can a bar graph showing the number of students failing a certain course in each of certain years be regarded as a function in elementary mathematics?

(f) In what sense can a table of trigonometric tangents be regarded as a function in elementary mathematics?

(g) Compare a function to a machine equipped with a hopper at the top and a spout at the bottom. Raw materials are fed into the machine through the hopper. The machine then converts these raw materials into a product which is expelled through the spout of the machine.

Further Reading

BAUM, J. D., *Elements of Point Set Topology*. Englewood Cliffs, N.J.: Prentice-Hall, 1964.

FREGE, GOTTLIEB, *The Foundations of Arithmetic*. Oxford: Basil Blackwell Mott, 1950.

HAUSDORFF, FELIX, *Set Theory*, tr. by J. R. Aumann. New York: Chelsea, 1957.

MENDELSON, BERT, *Introduction to Topology*, 2nd ed. Boston: Allyn and Bacon, 1968.

ROTMAN, B., and G. T. KNEEBONE, *The Theory of Sets and Transfinite Numbers*. London: Oldbourne, 1966.

RUSSELL, BERTRAND, *Introduction to Mathematical Philosophy*. London: George Allen and Unwin, 1956.

BEYOND THE FINITE

Mathematicians, and philosophers, have wrestled with the concepts of infinity and infinite sets from the days of the ancient Greeks, Zeno's paradoxes being an early indication of some of the difficulties that were encountered. Some of the Greeks, Aristotle and Proclus among them, accepted the fact that a set can be made larger and larger without bound, but denied the existence of a completed infinite set. Throughout the Middle Ages philosophers argued over this matter of the potential versus the actual infinite. It was noted that the comparison of certain infinite sets leads to paradoxes. For example, the points of two concentric circles can be put into one-to-one correspondence by associating points on a common radius; yet one circumference is longer than the other and hence would seem to contain more points than the other.

Galileo struggled with infinite sets, and he, too, felt that their completed actuality must be discarded. In his *Two New Sciences* of 1638, he noted that the points of two unequal line segments can be put into one-to-one correspondence by a simple projection of one segment onto the other, and so presumably the two segments contain the same number of points, though one segment is longer than the other and hence would seem to have more points than the other. He also noted that the positive integers can be put into one-to-one correspondence with their squares, by associating each positive integer with its square, though the set of squares of positive integers forms only a part of the set of all positive integers. These disturbing paradoxes arise only if one assumes the existence of completed infinite sets; to avoid the paradoxes one must reject the idea of a completed infinite set.

159

Gauss, in a famous letter to Schumacher dated July 12, 1831, says: "I protest against the use of an infinite quantity as an actual entity; this is never allowed in mathematics. The infinite is only a manner of speaking, in which one properly speaks of limits to which certain ratios can come as near as desired, while others are permitted to increase without bound." Cauchy, along with many others, similarly denied the existence of completed infinite sets on the paradoxical grounds that such a completed set can be put into one-to-one correspondence with a proper part of itself. Thus, though mathematicians worked with infinite sets, such as infinite series, the real numbers, the natural numbers, and so on, they generally avoided the troublesome problem behind the assumption that the completed sets exist. When mathematicians had finally to face the problem of instituting rigor into analysis, such matters could no longer be ignored or cursorily rejected.

Bernhard Bolzano (1781–1848), in his *Paradoxes of the Infinite,* published in 1851, three years after his death, was the first to take positive steps toward the acceptance of the actual existence of infinite sets. He said that the fact that an infinite set can be put into one-to-one correspondence with a proper part of itself must simply be accepted as a fact. But Bolzano's work on the infinite, though trailbreaking, was on the whole more philosophical than mathematical. A truly mathematical treatment of infinite sets did not occur until the remarkable work of Georg Cantor toward the end of the nineteenth century.

Georg Ferdinand Ludwig Philip Cantor was born of Danish parents in St. Petersburg, Russia, in 1845, and moved with his parents to Frankfurt, Germany, in 1856. Cantor's father was a Jew converted to Protestantism, and his mother had been born a Catholic. The son took a deep interest in medieval theology and its intricate arguments on the continuous and the infinite. As a consequence he gave up his father's suggestion of preparing for a career in engineering for concentrating on philosophy, physics, and mathematics. He studied at Zurich, Göttingen, and Berlin (where he came under the influence of Weierstrass and where he took his doctorate in 1869). He then spent a long teaching career at the University of Halle from 1869 until 1905. He died in a mental hospital in Halle in 1918.

Cantor's early interests were in number theory, indeterminate equations, and trigonometric series. The subtle theory of trigonometric series seems to have inspired him to look into the foundations of analysis. He produced his beautiful treatment of irrational numbers—which utilizes convergent sequences of rational numbers and differs radically from the geometrically inspired treatment of Dedekind—and commenced in 1874 his revolutionary work on set theory and the theory of the infinite. With this latter work, Cantor created a whole new field of mathematical research. In his papers, he developed a theory of transfinite numbers, based on a mathematical treatment of the actual infinite, and created an arithmetic of transfinite numbers analogous to the arithmetic of finite numbers.

Cantor was deeply religious, and his work, which in a sense is a continuation of the arguments connected with the paradoxes of Zeno, reflects his sympathetic respect for medieval scholastic speculation on the nature of the infinite. His views met considerable opposition, chiefly from Leopold Kronecker (1823–1891) of the University of Berlin, and it was Kronecker who steadfastly opposed Cantor's efforts toward securing a teaching post at the University of Berlin. Today, Cantor's set theory has penetrated into almost every branch of mathematics, and it has proved to be of particular importance in topology and the foundations of real function theory. There are logical difficulties, and paradoxes have appeared. The twentieth-century controversy between the formalists, led by Hilbert, and the intuitionists, led by Brouwer, is essentially a continuation of the controversy between Cantor and Kronecker. We shall look deeper into these matters in a later lecture.

In the previous lecture we saw that the cardinal numbers of finite sets can be identified with the natural numbers. The cardinal numbers of infinite sets are known as *transfinite numbers*. Cantor's development of the theory of these numbers appeared in a remarkable series of articles running from 1874 to 1895 and published, for the most part, in the German mathematics journals *Mathematische Annalen* and *Journal für Mathematik*.

Prior to Cantor's study, mathematicians had considered only one infinity, denoted by some symbol like ∞, and this symbol was employed indiscriminately to represent the "number" of elements in such sets as the set of all natural numbers, the set of all real numbers, and the set of all points on a given line segment. With Cantor's work, a whole new outlook was introduced, and a scale and arithmetic of infinities was achieved. Because of the unusual boldness of some of the ideas found in Cantor's work, and because of some of the singular methods of proof to which his work has given rise, Cantor's theory of transfinite numbers is indescribably fascinating. It certainly constitutes a GREAT MOMENT IN MATHEMATICS. Let us briefly examine this remarkable theory.

We employ the basic principle, given in the previous lecture, that equivalent sets are to bear the same cardinal number. This principle presents us with many interesting and intriguing situations when the sets under consideration are infinite sets. Thus Galileo's observation that, by the correspondence $n \leftrightarrow n^2$, the set of all squares of positive integers is equivalent to the set of all positive integers shows that the same cardinal number should be assigned to these two sets, and, from this point of view, we must say that there are as many squares of positive integers as there are positive integers in all. It follows that the Euclidean postulate which states that the whole is greater than a part does not hold when cardinal numbers of infinite sets are under consideration.

We shall designate the cardinal number of the set of all natural numbers by d,* and describe any set having this cardinal number as being *denumerable*. It follows that a set S is denumerable if and only if its elements can be written as an unending sequence $\{s_1, s_2, s_3, \ldots\}$. Since it is easily shown that any infinite set contains a denumerable subset, it follows that d is the "smallest" transfinite number.

Cantor, in one of his earliest papers on set theory, proved the denumerability of two important sets which scarcely seem at first glance to possess this property.

The first set is the set of all rational numbers. This set has the important property of being *dense*. By this is meant that between any two distinct rational numbers there exists another rational number—in fact, infinitely many other rational numbers. For example, between 0 and 1 lie the rational numbers

$$1/2, 2/3, 3/4, 4/5, 5/6, \ldots, n/(n + 1), \ldots;$$

---

*Cantor designated this cardinal number by the Hebrew letter aleph, with the subscript zero, that is, by $\aleph_0$.

between 0 and 1/2 lie the rational numbers

$$1/3, 2/5, 3/7, 4/9, 5/11, \ldots, n/(2n + 1), \ldots;$$

between 0 and 1/3 lie the rational numbers

$$1/4, 2/7, 3/10, 4/13, 5/16, \ldots, n/(3n + 1), \ldots;$$

and so on. Because of this property one might well expect the transfinite number of the set of all rational numbers to be greater than d.* Cantor showed that this is not the case and that, on the contrary, the set of all rational numbers is denumerable. His proof is interesting and runs as follows.

THEOREM 1. *The set of all rational numbers is denumerable.*

Consider the array

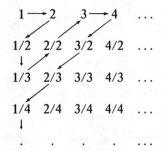

in which the first row contains, in order of magnitude, all the natural numbers (that is, all positive fractions with denominator 1); the second row contains, in order of magnitude, all the positive fractions with denominator 2; the third row contains, in order of magnitude, all the positive fractions with denominator 3; etc. Obviously, every positive rational number appears in this array, and if we list the numbers in the order of succession indicated by the arrows, omitting numbers which have already appeared, we obtain an unending sequence

---

*The cardinal number of a set A is said to be *greater than* the cardinal number of a set B if and only if B is equivalent to a proper subset of A, but A is not equivalent to any proper subset of B.

$$1, 2, 1/2, 1/3, 3, 4, 3/2, 2/3, 1/4, \ldots,$$

in which each positive rational number appears once and only once. Denote this sequence by $\{r_1, r_2, r_3, \ldots\}$. Then the sequence

$$0, -r_1, r_1, -r_2, r_2, \ldots$$

contains the set of all rational numbers, and the denumerability of the set is established.

The second set considered by Cantor is a seemingly much more extensive set of numbers than the set of rational numbers. We first make the following definition.

DEFINITION 1. A complex number is said to be *algebraic* if it is a root of some polynomial equation

$$f(x) = a_0 x^n + a_1 x^{n-1} + \cdots + a_{n-1} x + a_n = 0,$$

where $a_0 \neq 0$ and all the a_k's are integers. A complex number which is not algebraic is said to be *transcendental*.

It is quite clear that the algebraic numbers include, among others, all rational numbers and all roots of such numbers. Accordingly, the following theorem is somewhat astonishing.

THEOREM 2. *The set of all algebraic numbers is denumerable.*

Let $f(x)$ be a polynomial of the kind described in Definition 1, where, without loss of generality, we may suppose $a_0 > 0$. Consider the so-called *height* of the polynomial, defined by

$$h = n + a_0 + |a_1| + |a_2| + \cdots + |a_{n-1}| + |a_n|.$$

Obviously, h is an integer ≥ 1, and there are plainly only a finite number of polynomials of a given height h, and therefore only a finite number of algebraic numbers arising from polynomials of a given height h. We may now list (theoretically speaking) all the algebraic numbers, refraining from repeating any number already listed by first taking those arising from polynomials of height 1, then those arising from polynomials of height 2, then those arising from polynomials of height 3, and so on. We thus see that the set of all algebraic numbers can be listed in an unending sequence; hence the set is denumerable.

In view of the past two theorems, there remains the possibility that all infinite sets are denumerable. That this is not so was shown by Cantor in a striking proof of the following significant theorem.

THEOREM 3. *The set of all real numbers in the interval* $0 < x < 1$ *is nondenumerable.*

The proof is indirect and employs an unusual method known as the *Cantor diagonal process.* Let us, then, assume the set to be denumerable. Then we may list the numbers of the set in a sequence $\{p_1, p_2, p_3, \ldots\}$. Each of these numbers p_i can be written uniquely as a nonterminating decimal fraction; in this connection it is useful to recall that every rational number may be written as a "repeating decimal"; a number such as 0.3, for example, can be written as 2.9999 .... We can then display the sequence in the following array,

$$p_1 = 0.a_{11}a_{12}a_{13}\ldots$$
$$p_2 = 0.a_{21}a_{22}a_{23}\ldots$$
$$p_3 = 0.a_{31}a_{32}a_{33}\ldots$$

$$\cdots\cdots\cdots,$$

where each symbol a_{ij} represents some one of the digits 0, 1, 2, 3, 4, 5, 6, 7, 8, 9. Now, in spite of any care which has been taken to list all the real numbers between 0 and 1, there is a number which could not have been listed. Such a number is $0.b_1b_2b_3\ldots$, where, say, $b_k = 7$ if $a_{kk} \neq 7$ and $b_k = 3$ if $a_{kk} = 7$, for $k = 1, 2, 3, \ldots, n, \ldots$. This number clearly lies between 0 and 1, and it must differ from each number p_i, for it differs from p_1 in at least the first decimal place, from p_2 in at least the second decimal place, from p_3 in at least the third decimal place, and so on. Thus the original assumption that all the real numbers between 0 and 1 can be listed in a sequence is untenable, and the set must therefore be nondenumerable.

Cantor deduced the following remarkable consequence of Theorems 2 and 3.

THEOREM 4. *Transcendental numbers exist.*

Since, by Theorem 3, the set of all real numbers between 0 and 1 is nondenumerable, it certainly follows that the set of all complex numbers is also nondenumerable. But, by Theorem 2, the set of all algebraic numbers is denumerable. It follows that there must exist complex numbers which are not algebraic, and the theorem is established.

Not all mathematicians are willing to accept the proofs above of Theorems 3 and 4. The acceptability or nonacceptability of the proofs hinges upon what one believes mathematical existence to be, and there are some mathematicians who feel that mathematical existence is established only when one of the objects whose existence is in question is actually constructed and exhibited. Now the proof of Theorem 4 above, for example, does not establish the existence of transcendental numbers by producing a specific example of such a number. There are many existence proofs in mathematics of this nonconstructive sort, where existence is presumably established by merely showing that the assumption of nonexistence leads to a contradiction.* Because of the dissatisfaction of some mathematicians with nonconstructive existence proofs, a good deal of effort has been made to replace such proofs by those which actually yield one of the objects concerned.

The proof of the existence of transcendental numbers and the proof that some particular number is transcendental are two quite different matters, the latter often being a very difficult problem. It was Charles Hermite (1822-1901) who, in 1873, proved that the number e, the base for natural logarithms, is transcendental, and C. L. F. Lindemann (1852-1939), in 1882, who first established the transcendentality of the number π. Unfortunately, it is inconvenient for us to prove these interesting facts here, but, in a more extended listing, each of these accomplishments would be considered as a GREAT MOMENT IN MATHEMATICS. The difficulty of identifying a particular given number as algebraic or transcendental is illustrated by the fact that it is not yet known whether the number π^π is algebraic or transcendental. A recent gain along these lines was the establishment of the transcendental character of any number of the form a^b, where a is an algebraic number different from 0 and 1, and b is any irrational algebraic number. This result was a culmination of an almost

---

*A constructive proof of the existence of transcendental numbers was given by Joseph Liouville (1809-1882) in 1851, twenty-some years before Cantor published his nonconstructive demonstration.

30-years' effort to prove that the so-called *Hilbert number,* $2^{\sqrt{2}}$, is transcendental.

Since the set of all real numbers in the interval $0 < x < 1$ is non-denumerable, the transfinite number of this set is greater than d. We shall denote it by c. It is generally believed that c is the next trans-finite number after d, that is, that there is no set having a cardinal number greater than d but less than c. This belief is known as the *continuum hypothesis,* but, in spite of strenuous efforts, no proof has been found to establish or disestablish it. Many consequences of the hypothesis have been deduced, and, about 1940, the Austrian logician Kurt Gödel (1906–1978) succeeded in showing that the con-tinuum hypothesis is consistent with a famous postulate set of set theory provided this postulate set itself is consistent. Gödel conjec-tured that the denial of the continuum hypothesis is also consistent with the postulates of set theory. This conjecture was established, in 1963, by Paul J. Cohen of Stanford University, thus proving that the continuum hypothesis is independent of the postulates of set theory, and hence can never be deduced from those postulates. The situation is analogous to that of the parallel postulate in Euclidean geometry.

It has been shown that the set of all single-valued functions $f(x)$ defined over the interval $0 < x < 1$ has a cardinal number greater than c, but whether this cardinal number is or is not the next after c is not known. Cantor's theory provides for an infinite sequence of transfinite numbers, and there are demonstrations which purport to show that an unlimited number of cardinal numbers greater than c actually exist.

Exercises

34.1. (a) Prove that the union of a finite number of denumerable sets is a denumerable set.

(b) Prove that the union of a denumerable number of denumera-ble sets is a denumerable set.

(c) Show that the set of all irrational numbers is nondenumerable.

(d) Show that the set of all transcendental numbers is nonde-numerable.

34.2. (a) Show that 1 is the only polynomial of height 1.

(b) Show that x and 2 are the only polynomials of height 2.

(c) Show that x^2, $2x$, $x + 1$, $x - 1$, and 3 are the only polynomials of height 3, and they yield the distinct algebraic numbers 0, 1, -1.

(d) Form all possible polynomials of height 4 and show that the only new real algebraic numbers contributed are -2, $-1/2$, $1/2$, 2.

(e) Show that polynomials of height 5 contribute 12 more real algebraic numbers.

34.3. (a) Complete the details of the following proof that the set of all points on a line segment AB is nondenumerable.

Take the length of AB to be 1 unit, and assume that the points on AB constitute a denumerable set. The points on AB can then be arranged in a sequence $\{P_1, P_2, P_3, \ldots\}$. Enclose point P_1 in an interval of length $1/10$, point P_2 in an interval of length $1/10^2$, point P_3 in an interval of length $1/10^3$, and so on. It follows that the unit interval AB is entirely covered by an infinite sequence of possibly overlapping subintervals of lengths $1/10$, $1/10^2$, $1/10^3$, $\ldots$. But the sum of the lengths of these subintervals is

$$1/10 + 1/10^2 + 1/10^3 + \cdots = 1/9 < 1.$$

(b) By choosing the subintervals in part (a) to be of lengths $\epsilon/10$, $\epsilon/10^2$, $\epsilon/10^3$, $\ldots$, where ϵ is an arbitrarily small positive number, show that a denumerable set of points can be covered by a set of intervals the sum of whose lengths can be made as small as we please. (Using the terminology of measure theory, we say that a denumerable set of points has *zero measure*.)

34.4. Let E_1 denote the set of all points on a unit line segment, and let E_2 denote the set of all points in a unit square. A point Z of E_1 may be designated by an unending decimal $z = 0.z_1z_2z_3\ldots$ lying between 0 and 1, and a point P of E_2 may be designated by an ordered pair of unending decimals

$$(x = 0.x_1x_2x_3\ldots, y = 0.y_1y_2y_3\ldots),$$

each decimal lying between 0 and 1. Suppose we let z_i, x_i, y_i in these representations denote either a nonzero digit or a nonzero digit preceded by a possible block of zeros. For example, if $z = 0.73028007\ldots$, then $z_1 = 7$, $z_2 = 3$, $z_3 = 02$, $z_4 = 8$, $z_5 = 007$, $\ldots$. Show that a one-to-one correspondence may be set up between the points of E_1 and those of E_2 by associating with the point $0.z_1z_2z_3\ldots$ of E_1 the point

$$(0.z_1z_3z_5\ldots, 0.z_2z_4z_6\ldots)$$

of E_2, and the point

$$(0.x_1x_2x_3\ldots, 0.y_1y_2y_3\ldots)$$

of E_2 the point $0.x_1y_1x_2y_2x_3y_3\ldots$ of E_1. Thus show that the set of all points in a unit square has the transfinite number c. (This shows that the dimension of a manifold cannot be distinguished by the transfinite number of the manifold.)

34.5. (a) Show that if a circle has a center with at least one irrational coordinate, then there are at most two points on the circle with rational coordinates.

(b) Show that if a circle has a center with at least one transcendental coordinate, then there are at most two points on the circle with algebraic coordinates.

(c) Is it possible for a straight line or a circle in the Cartesian plane to contain only points having rational coordinates? Algebraic coordinates?

(d) Show that any infinite set of mutually external closed intervals on a straight line is denumerable.

(e) Show that any infinite set of mutually external circles lying in a plane is denumerable.

34.6. (a) Show that every rational number is an algebraic number and hence that every real transcendental number is irrational.

(b) Is every irrational number a transcendental number?

(c) Is the imaginary unit i algebraic or transcendental?

(d) Show that $\pi/2$ is transcendental.

(e) Show that $\pi + 1$ is transcendental.

(f) Generalize parts (d) and (e).

(g) Show that any complex number which is a zero of a polynomial of the form

$$a_0x^n + a_1x^{n-1} + \cdots + a_{n-1}x + a_n,$$

where $a_0 \neq 0$ and all the a_k's are *rational numbers,* is an algebraic number.

34.7. Show that the cancellation laws for addition and multiplication do not hold for transfinite numbers.

34.8. Show that the set of all finite sequences of nonnegative integers is denumerable.

34.9. Let S denote the set of all single-valued functions of one variable x which assume positive integral values whenever x is a positive integer. Show that S is nondenumerable.

Further Reading

BELL, E. T., *Men of Mathematics.* New York: Simon and Schuster, 1937.

BOLZANO, BERNHARD, *Paradoxes of the Infinite,* tr. by F. Prihonsky. London: Routledge and Kegan Paul, 1950.

BREUER, JOSEPH, *Introduction to the Theory of Sets,* tr. by H. F. Fehr. Englewood Cliffs, N.J.: Prentice-Hall, 1958.

CANTOR, GEORG, *Contributions to the Founding of the Theory of Transfinite Numbers,* tr. by P. E. B. Jourdain. New York: Dover, 1915.

GALILEO GALILEI, *Dialogue Concerning Two New Sciences,* tr. by H. Crew and A. de Salvio. New York: Dover, 1951.

JOHNSON, P. E., *A History of Set Theory.* Boston: Prindle, Weber & Schmidt, 1972.

KAMKE, E., *Theory of Sets,* tr. by F. Bagemihl. New York: Dover, 1950.

ROTMAN, B., and G. T. KNEEBONE, *The Theory of Sets and Transfinite Numbers.* London: Oldbourne, 1966.

SOME REMARKABLE DEFINITIONS

Though the ancient Greeks contributed enormously to the content of mathematics, perhaps their most outstanding contribution to the subject was their organization of mathematics by the axiomatic method. This earliest form of the axiomatic method has become known as *material axiomatics,* and a description of its pattern was given in LECTURE 7. In LECTURE 8 we considered Euclid's *Elements,* which is the first great application of material axiomatics that has come down to us.

The Greek concept of axiomatics persisted, after a period of general neglect, into the nineteenth century, when three cardinal events in the development of mathematics led mathematicians to a deeper study of axiomatic procedure. The first two of these events were the discovery, about 1829, of a non-Euclidean geometry, and, not long after, the discovery of a noncommutative algebra. These two discoveries cast skepticism on the early Greek concept of an axiom as a self-evident, or at least readily acceptable, truth. The third event was the long examination that culminated in the arithmetization of analysis. In constructing suitable postulational bases for the real number system and the natural number system, the whole axiomatic method received a very careful scrutiny.

With the work of a number of nineteenth-century mathematicians, the axiomatic method was gradually sharpened and refined. It was Hilbert's *Grundlagen der Geometrie,* which was a popular and rigorous postulational development of Euclidean geometry and which appeared in 1899, that proved most influential in this revamping of the axiomatic method. From the material axiomatics of the Greeks evolved the so-called *formal axiomatics* of the early twentieth century.

171

To help clarify the difference between the two forms of axiomatics let us introduce the modern concept of *propositional function*, the fundamental importance of which was first brought to notice by the English mathematician and philosopher Bertrand Russell (1872–1970).

Consider the three statements:

(1) Spring is a season.
(2) 8 is a prime number.
(3) x is a y.

Each of these statements has form—the same form; each asserts that "something" is a "something else." Statements (1) and (2) have content as well as form; statement (3) has form only. Knowing what is meant by the concrete terms "spring" and "season," we claim that statement (1) is a *true* statement, and knowing what is meant by the concrete terms "8" and "prime number," we claim that statement (2) is a *false* statement. It follows that statements (1) and (2) are propositions, for, by definition, a proposition is any statement to which it is meaningful to assign the quality of truth or of falseness. Statement (3), on the other hand, since it asserts nothing definite, is neither true nor false, and therefore is *not* a proposition. Statement (3), however, though not a proposition, does have the form of a proposition. It has been called a *propositional function*, for if in the form

$$x \text{ is a } y$$

we substitute terms of definite meaning for the variables x and y, we obtain propositions, true propositions if the substituted terms should verify the propositional function, false propositions if the substituted terms should falsify the propositional function. The form considered above is a propositional function in two variables, and there are infinitely many verifying values for the two variables.

A propositional function may contain any number of variables. An example having but one is: x is a volume in the Library of Congress. Here x evidently has as many verifying values as there are volumes in the Library of Congress. Evidently, too, the variable x has many falsifying values.

There is no need for the variables in a propositional function to be denoted by symbols, such as x, y, ...; they may be ordinary words.

Thus, should a statement whose terms are ordinary words appear in a discourse with no indication as to the senses in which the words are to be understood, then in that discourse the statement is really a propositional function, rather than a proposition, and in the interests of clarity the ambiguous or undefined terms might better be replaced by such symbols as x, y, ... Written and spoken discourse often contains such statements, and though asserted by their authors to be propositions, true or false, are in reality propositional functions, devoid of all true or false quality. This fact perhaps accounts for much of the argument and misunderstanding among people.

With the idea of a propositional function firmly in mind, let us return to our discussion of axiomatic procedure. We recall, from LECTURE 7, that any logical discourse, in an endeavor to be clear, tries to define explicitly the elements of the discourse, the relations among these elements, and the operations to be performed upon them. Such definitions, however, must employ other elements, relations, and operations, and these, too, are subject to explicit definition. If these are defined, it must again be by reference to further elements, relations, and operations. There are two roads open to us; either the chain of definitions must be cut short at some point or else it must be circular. Since circularity is not to be tolerated in a logical discourse, the definitions must be brought to a close at some point; thus it is necessary that one or more elements, relations, and/or operations receive no explicit definition. These are known as the *primitive terms* of the discourse.

There is also, in a logical discourse, an effort to deduce the statements of the discourse. Such deductions, however, are from other statements, and these other statements, too, must be deduced from some still further statements. Again, in order to get started and also to avoid the vicious circle, one or more of the statements must remain entirely *unproved*. These are known as the *postulates* (or axioms, or primary statements) of the discourse.

Clearly, then, any logical discourse such as we are considering must conform to the following pattern.

Pattern of Formal Axiomatics

(A) The discourse contains a set of technical terms (elements, relations among elements, operations to be performed on elements) which are deliberately chosen as undefined terms. These are the *primitive terms* of the discourse.

(B) The discourse contains a set of statements about the primitive terms which are deliberately chosen as unproved statements. These are called the *postulates* (or *axioms*), P, of the discourse. Parts (A) and (B) are jointly referred to as the *basis* of the discourse.

(C) All other technical terms of the discourse are defined by means of previously introduced terms.

(D) All other statements of the discourse are logically deduced from previously accepted or established statements. These derived statements are called the *theorems, T,* of the discourse.

(E) For each theorem T_i of the discourse there exists a corresponding statement (which may or may not be formally expressed) asserting that theorem T_i is logically implied by the postulates P. (Often a corresponding statement appears at the end of the proof of the theorem in some such words as, "Hence the theorem ... " or "This completes the proof of the theorem." In some elementary geometry textbooks the statement appears, at the end of the proof of the theorem, as "Q.E.D." (*Quod Erat Demonstrandum*). The modern symbol ■, or some variant of it, suggested by Paul R. Halmos, is frequently used to signal the end of a proof.)

The first thing to notice in the pattern above is that the primitive terms, being undefined terms, might just as well (if such is not already the case) be replaced by symbols like $x, y, \ldots$. Let us suppose this substitution is made. Then the primitive terms are clearly variables. The second thing to notice is that the postulates, P, since they are statements about the primitive terms, are nothing more than propositional functions and are thus devoid of all true or false content. The third thing to notice is that the theorems, T, since they are but logical implications of the postulates, P, also are propositional functions and thus devoid of all true or false content.

Since the postulates and the theorems of a logical discourse are merely propositional functions, that is, are statements of form only and without content, it would seem that the whole discourse is somewhat vacuous and entirely bereft of truth or falseness. Such, however, is not the case, for by (E) of the postulational pattern, we have the all-important statement,

(F) The postulates P imply the theorems T.

Now (F) asserts something definite; it is true or false, and so is a proposition—a true one if the theorems T are in fact implied by the postulates P, and a false one if they are not. The statement (F) is precisely what the discourse is designed for; it is the discourse's sole aim and excuse for being.

We come now to some remarkable definitions. A discourse conducted according to the pattern above has been called, by some mathematicians, a *branch of pure mathematics,* and the grand total of all such existing branches of pure mathematics, the *pure mathematics of to-date.*

If, for the variables (the primitive terms) in a branch of pure mathematics we should substitute terms of definite meaning which convert all the postulates of the branch into true propositions, then the set of substituted terms is called an *interpretation* of the branch of pure mathematics. The interpretation will also, provided all deductions have been correctly performed, convert the theorems of the discourse into true propositions. The result of substituting an interpretation into a branch of pure mathematics is called a *model* of the branch of pure mathematics.

A model of a branch of pure mathematics has been called a *branch of applied mathematics,* and the grand total of all existing branches of applied mathematics, the *applied mathematics of to-date.* Thus the difference between applied and pure mathematics is not one of applicability and inapplicability, but rather of concreteness and abstractness. Behind every branch of applied mathematics lies a branch of pure mathematics, the latter being an abstract development of what formerly was a concrete development. It is conceivable (and indeed such is often the case) that a single branch of pure mathematics may have several models, or associated branches of applied mathematics. This is the "economy" feature of pure mathematics, for the establishment of a branch of pure mathematics automatically assures the simultaneous establishment of all its allied branches of applied mathematics.

The abstract development of some branch of pure mathematics is an instance of *formal axiomatics,* whereas the concrete development of a given branch of applied mathematics is an instance of *material axiomatics.* In the former case we think of the postulates as prior to any specification of the primitive terms, and in the latter we think of the objects that interpret the primitive terms as being prior to the postulates. In the former case a postulate is simply a basic assumption about some undefined primitive terms; in the latter case a postulate expresses some property of the basic objects which is taken as initially evident or acceptable. This latter is the older view of a postulate, and was the view held by the ancient Greeks. Thus, to the Greeks, geometry was thought of as a study dealing with a unique structure of physical space, in which the elements *points* and *lines* are regarded as idealizations of certain actual physical entities, and in which the postulates are readily accepted statements about these idealizations. From the modern point of view, geometry is a purely abstract study devoid of any physical meaning or imagery.

The notion of pure mathematics gives considerable sense to Bertrand Russell's seemingly facetious remark that "mathematics may be defined as the subject in which we never know what we are talking about, nor whether what we are saying is true." It also accords with Henri Poincaré's (1854–1912) saying that mathematics is "the giving of the same name to different things," and with Benjamin Peirce's (1809–1880) remark that "mathematics is the science which draws necessary conclusions."

Not all mathematicians subscribe to the definition above of (pure) mathematics, though one can cite many corroborating statements, similar to those just given, that have been made by other eminent mathematicians. There are many mathematicians, practicing what they regard to be mathematics, who never work with postulate sets; nevertheless, they are usually deducing consequences of certain given or assumed premises, and this is largely the same thing. Except for beginning geometry, practically no other lower school or early college mathematics course is developed via the axiomatic method. However, these other subjects can be so developed; it is merely that to do so at the elementary level is perhaps too sophisticated for the student, and therefore ill-advised on pedagogical grounds. Moreover, it must be admitted that much that is done at the elementary level is really not so much mathematics as only the mastering of computational techniques. In any event, the evolution of formal axiomatics constitutes an outstanding GREAT MOMENT IN MATHEMATICS, as does the achievement of the definition above, as one possible definition, of (pure) mathematics. Indeed, Bertrand

Russell regarded it as a very high and remarkable achievement that in the twentieth century mathematicians finally realized precisely what mathematics is.

We shall conclude our lecture with a word or two as to how postulate sets in mathematics arise. Most postulate sets originate with a model of some abstract system. That is, in most cases, the individual researcher has in mind some specific model, and then proceeds to build a postulate system that will recover his model, so that, in practice, the model or interpretation generally comes first and the postulate set later. This is particularly true in such fields as biology, chemistry, economics, ethics, law, mechanics, philosophy, physics, zoology, and so on, where someone familiar with the field might choose to set down some postulates for the field, or for some part of the field, and then see what theorems can be logically deduced from the postulates. Theoretically, of course, there is no need for the model to precede the abstract development, though it must be admitted that this is the usual source of a postulate set. If one were to put down some arbitrary symbols for primitive terms and then try to formulate postulates about these terms, it would be difficult, without some model in mind, to think of anything to say, and, moreover, one would have to take great care to ensure consistency of the postulates. Sometimes a new postulate system is derived from a given postulate system by altering one or more postulates of the given system. It was in this way that postulate sets for Lobachevskian and Riemannian non-Euclidean geometries were derived from a postulate set for Euclidean geometry. Other examples can be given.

Exercises

35.1. (a) Construct an example of a propositional function containing three variables.

(b) Obtain, by appropriate substitutions, three true propositions from the propositional function.

(c) Obtain, by appropriate substitutions, three false propositions from the propositional function.

35.2. Discuss the statement, "God is love," in terms of the concepts *proposition* and *propositional function*.

35.3. In elementary mathematics, an equation in one variable x, say, can be regarded as a propositional function in which the only admissible meanings for x are complex numbers. A fundamental problem in elementary mathematics is the determination of those complex numbers, if there are such, for which the equation becomes a true proposition. What are the verifying values of x for each of the following equations?

(a) $3150 + 15x = 3456 - 33/2x$.
(b) $\sin x = \cos x$.
(c) $2x^2 - 5x - 3 = 0$.
(d) $x^2 = 3 - x$.
(e) $x^2 + 3x + 6 = 0$.
(f) $2x/(x^2 - 9) + 2/(x + 3) = 1/(x - 3)$.
(g) $\sqrt{x + 2} + \sqrt{x^2 - 4} = 0$.

35.4. One purpose of formal axiomatics, with its abstraction and symbolism, is to furnish a protective cloak against the use of intuition. Answer the following questions intuitively, and then check your answers by calculation.

(a) A car travels from P to Q at the rate of 40 miles per hour and then returns from Q to P at the rate of 60 miles per hour. What is the average rate for the round trip?

(b) A can do a job in 4 days, and B can do it in 6 days. How long will it take A and B together to do the job?

(c) A man sells half of his apples at 3 for 17 cents and then sells the other half at 5 for 17 cents. At what rate should he sell all his apples to make the same income?

(d) If a ball of yarn 4 inches in diameter costs 20 cents, how much should you pay for a ball of yarn 6 inches in diameter?

(e) Two jobs have the same starting salary of $6000 per year and the same maximum salary of $12,000 per year. One job offers an annual raise of $800 and the other offers a semiannual raise of $200. Which is the better paying job?

(f) Each bacterium in a certain culture divides into two bacteria once a minute. If there are 20 million bacteria present at the end of one hour, when were there exactly 10 million bacteria present?

(g) Is a salary of 1 cent for the first half month, 2 cents for the second half month, 4 cents for the third half month, 8 cents for the fourth half month, and so on until the year is used up, a good or poor total salary for the year?

(h) A clock strikes six in 5 seconds. How long will it take to strike twelve?

(i) A bottle and a cork together cost $1.10. If the bottle costs a dollar more than the cork, how much does the cork cost?

(j) Suppose that in one glass there is a certain quantity of a liquid A, and in a second glass an equal quantity of another liquid, B. A spoonful of liquid A is taken from the first glass and put into the second glass, then a spoonful of the mixture from the second glass is put back into the first glass. Is there now more or less liquid A in the second glass than there is liquid B in the first glass?

(k) Suppose that a large sheet of paper one one-thousandth of an inch thick is torn in half and the two pieces put together, one on top of the other. These are then torn in half, and the four pieces put together in a pile. If this process of tearing in half and piling is done 50 times, will the final pile of paper be more or less than a mile high?

(l) Is a discount of 15 percent on the selling price of an article the same as a discount of 10 percent on the selling price followed by a discount of 5 percent on the reduced price?

(m) Four-fourths exceeds three-fourths by what fractional part?

(n) A boy wants the arithmetic average of his eight grades. He averages the first four grades, then the last four grades, and then finds the average of these averages. Is this correct?

35.5. Though the primitive terms of a discourse conducted by formal axiomatics receive no *explicit* definition, discuss how they may be said to receive *implicit* definition.

35.6. In each of the following, is the given conclusion a valid deduction from the given pair of premises?

(a) If today is Saturday, then tomorrow will be Sunday.
   But tomorrow will be Sunday.
   Therefore, today is Saturday.

(b) Germans are heavy drinkers.
   Germans are Europeans.
   Therefore, Europeans are heavy drinkers.

(c) If a is b, then c is d.

But c is d.

Therefore, a is b.

(d) All a's are b's.

All a's are c's.

Therefore, all c's are b's.

These exercises illustrate how a person may allow the meanings which he associates with words or expressions to dominate his logical analysis. There is a greater tendency to go wrong in (a) and (b) than in (c) and (d), which are symbolic counterparts of (a) and (b).

Further Reading

BLANCHÉ, ROBERT, *Axiomatics*, tr. by G. B. Keene. London: Routledge and Kegan Paul, 1962.

MEYER, BURNETT, *An Introduction to Axiomatic Systems*. Boston: Prindle, Weber & Schmidt, 1974.

WILDER, R. L., *Introduction to the Foundations of Mathematics*, 2nd ed. New York: John Wiley, 1965.

SOME CLARIFYING EXAMPLES

In this lecture we shall clarify, by means of examples, the interesting definitions encountered in the previous lecture. In short, we propose to give a simple example of a branch of pure mathematics followed by three applications of that branch. That is, we shall develop a short discourse by formal axiomatics and then, by appropriate interpretations of the primitive terms of the discourse, obtain three models of the discourse. We first explain some notation that will be employed.

Everyone is familiar with the idea of a dyadic relation as a form of connectivity between a pair of objects, for this is a concept which is not peculiar to mathematics but permeates everyday life and conversation. Thus it is common to hear of such dyadic relations as "is the father of," "is married to," "is to the left of," "is taller than," "is heavier than," "is the same color as," and so on. Elementary geometry is interested in such dyadic relations as "is similar to," "is congruent to," "is parallel to," and "is perpendicular to," and beginning arithmetic is concerned with such dyadic relations as "is equal to," "is not less than," and "is greater than." These phrases of relationship, or symbols for them, are usually placed between the objects so related, as "*a is the father of b*" and "*a is taller than b*," where *a* and *b* are people; "*a ~ b*" and "*a ≅ b*," where *a* and *b* are triangles, "*a ∥ b*" and "*a ⊥ b,*" where *a* and *b* are lines; "*a = b*" and *a ≮ b*" where *a* and *b* are real numbers. Accordingly, if two elements *a* and *b* of a set *K* are related by some dyadic relation *R*, we shall write "*aRb*" and read this as "*a is R-related to b*." If *a* is not *R*-related to *b* we shall write "*a̸Rb*." Clearly, if *aRb* it does not necessarily follow that *bRa* (consider, for example, the dyadic relation "is the father of" in connection with the set of all people).

In our work we shall use the equal sign in the sense of identity. Thus "$a = b$" will mean a and b are the same element. If a and b are distinct elements, we shall write "$a \neq b$."

We are now ready to give an example of a discourse conducted by formal axiomatics.

A Simple Example of a Branch of Pure Mathematics

Basis. A set K of undefined elements $a, b, c, \ldots$ and an undefined dyadic relation R connecting certain pairs of elements of K, satisfying the following four postulates:

P1. *If $a \neq b$, then either aRb or bRa.*

P2. *If aRb, then $a \neq b$.*

P3. *If aRb and bRc, then aRc.*

P4. *K consists of exactly four distinct elements.*

From the postulates above we shall now deduce some further statements, or *theorems*. These theorems will be designated by T1, T2, ... and any formulated definitions by D1, D2, ...

T1. *If aRb, then bRa.*

Suppose both *aRb and bRa*. Then, by P3, *aRa*. But this is impossible by P2. Hence the theorem by *reductio ad absurdum*.

T2. *If aRb, and c is in K, then either aRc or cRb.*

If $c = a$, then cRb, and we are done. If $c \neq a$, we have, by P1, either *aRc* or *cRa*. If *cRa*, since also *aRb*, we have, by P3, *cRb*. Hence the theorem.

T3. *There is at least one element of K not R-related to any element of K.*

Suppose the contrary case and let a be any element of K. Then, by our supposition, there exists an element b of K such that *aRb*. By P2, a and b are distinct elements of K.

By our supposition there exists an element c of K such that *bRc*.

By P2, $b \neq c$. By P3, we also have aRc. By P2, $a \neq c$. Thus a, b, c are distinct elements of K.

By our assumption there exists an element d of K such that cRd. By P2, $c \neq d$. By P3, we also have bRd and aRd. By P2, $b \neq d$, $a \neq d$. Thus a, b, c, d are distinct elements of K.

By our supposition there exists an element e of K such that dRe. By P2, $d \neq e$. By P3, we also have cRe, bRe, aRe. By P2, $c \neq e$, $b \neq e$, $a \neq e$. Thus a, b, c, d, e are distinct elements of K.

We now have a contradiction of P4. Hence the theorem by *reductio ad absurdum*.

T4. *There is only one element of K not R-related to any element of K.*

By T3, there is at least one such element, say a. Let $b \neq a$ be any other element of K. By P1, either aRb or bRa. But, by hypothesis, we do not have aRb. Therefore we must have bRa, and the theorem is proved.

D1. If bRa, then we say aDb.

T5. *If aDb and bDc, then aDc.*

By D1, bRa and cRb. By P3, we then have cRa, or, by D1, aDc.

D2. If aRb and there is no element c of K such that aRc and cRb, then we say aFb.

T6. *If aFc and bFc, then $a = b$.*

Suppose $a \neq b$. Then, by P1, either aRb or bRa.

Case 1. Suppose aRb. Since bFc, by D2, bRc. But this is impossible, since aFc.

Case 2. Suppose bRa. Since aFc, by D2, aRc. But this is impossible, since bFc.

Thus in either case we are led to a contradiction of our hypothesis, and the theorem follows by *reductio ad absurdum*.

T7. *If aFb and bFc, then $a\!\!\not\!Fc$.*

By D2, aRb and bRc. Hence, again by D2, $a\!\!\not\!Fc$.

D3. If aFb and bFc, then we say aGc.

We shall cut short our abstract postulational discourse at this point. There are, of course, many other theorems that can be established in the system, but probably we have considered enough to illustrate the notion of a branch of pure mathematics.

Before passing on to a clarification of further definitions encountered in the previous lecture, let us pause to make some remarks about this branch of pure mathematics. It will be noticed that the proof of the very first theorem, T1, was accomplished by employing *reductio ad absurdum*. Some of the other proofs were similarly established. The indirect method of proof is frequently used in the early part of an axiomatic development. This is because early in a deductive treatment there are so few statements available for direct use. Theorem T3 is an example of an *existence proof*; existence proofs are common and important in many bodies of mathematics. Theorem T4 is an example of a *uniqueness proof*; such proofs also are common and important in many bodies of mathematics. Sometimes it is expedient to establish a theorem by cases; the proof of theorem T6 is an example of this.

If for every pair of distinct elements a and b of a set K we have, for a dyadic relation R, either aRb or bRa, we say that R is *determinate* in K. If aRa for every element a of K, we say that R is *reflexive* in K; if $a\not Ra$ for every element a of K, we say that R is *irreflexive* in K. If aRb implies bRa, we say R is *symmetric* in K; if aRb implies $b\not Ra$, we say R is *asymmetric* in K. If aRb and bRc imply aRc, we say R is *transitive* in K; if aRb and bRc imply $a\not Rc$, we say R is *intransitive* in K. With these notions about a dyadic relation it is worth noting that a number of the postulates and theorems of our branch of pure mathematics can be given more succinct statements. Thus P1, P2, P3, T1, T5, and T7 can be restated as:

P1. *R is determinate in K.*

P2. *R is irreflexive in K.*

P3. *R is transitive in K.*

T1. *R is asymmetric in K.*

T5. *D is transitive in K.*

T7. *F is intransitive in K.*

We now proceed to give three interpretations of our branch of pure mathematics, thus obtaining three models or three derived branches of applied mathematics.

Application 1 (Genealogical)

Let the elements of K be four men—some man, his father, his father's father, and his father's father's father, and let R mean "is an ancestor of."

We readily see that this interpretation of the elements of K and of the relation R converts the postulates into true propositions. We are thus led to a concrete model of our abstract postulational discourse, that is, to a branch of applied mathematics derived from our branch of pure mathematics. The theorems, which now must all become true propositions, and the definitions read:

T1(1). If a is an ancestor of b, then b is not an ancestor of a.

T2(1). If a is an ancestor of b, and if c is one of the four men, then either a is an ancestor of c or c is an ancestor of b.

T3(1). There is at least one man in K who is not an ancestor of anyone in K.

T4(1). There is only one man in K who is not an ancestor of anyone in K.

D1(1). If b is an ancestor of a, we say that a is a *descendant* of b.

T5(1). If a is a descendant of b and b is a descendant of c, then a is a descendant of c.

D2(1). If a is an ancestor of b and there is no individual c of K such that a is an ancestor of c and c is an ancestor of b, we say that a is a *father* of b.

T6(1). A man has at most one father.

T7(1). If a is the father of b and b is the father of c, then a is not the father of c.

D3(1). If a is the father of b and b is the father of c, we say that a is a *grandfather* of c.

The notations *D, F,* and *G* in our branch of pure mathematics show how that branch might have been obtained as an abstraction of this branch of applied mathematics. Many branches of pure mathematics arise in this way—as abstractions of some concrete model of the branch.

Application 2 (Geometrical)

Let the elements of *K* be four distinct points on a horizontal straight line, and let *R* mean "is to the left of."

Again our postulates are satisfied and we have a second branch of applied mathematics derived from our branch of pure mathematics. The relation *D* means "is to the right of," the relation *F* means "is the first point of *K* to the left of," and relation *G* means "is the second point of *K* to the left of."

Application 3 (Arithmetical)

Let the elements of *K* be the four integers 1, 2, 3, 4, and let *R* mean "is less than."

Once again our postulates are satisfied and we have a third branch of applied mathematics derived from our branch of pure mathematics. Here relation *D* means "is greater than," relation *F* means, "is 1 less than," and relation *G* means "is 2 less than."

Our example of a branch of pure mathematics, with its derived branches of applied mathematics, illustrates the "economy" feature of the modern axiomatic method. Any theorem of the branch of pure mathematics yields a corresponding theorem in each of the applications, and these latter require no proof so long as the theorem in the abstract system has been proved. The abstract postulate set studied above is a postulate set for *simple order* among four elements. Any relation which is an intepretation of the first three postulates* is called a *simple order relation.* "Is less than," "is greater than," "is

---

*These postulates were first studied by E. V. Huntington in 1905. See E. V. Huntington, *The Continuum and Other Types of Serial Order,* 2nd ed. New York: Dover, 1955.

to the left of," "is to the right of," "is before," "is after," are all simple order relations.

Exercises

36.1. Establish the following theorem of the simple branch of pure mathematics developed in the lecture text: *If aGc and bGc, then $a = b$.*

36.2. (a) In the simple branch of pure mathematics developed in the lecture text, define the *triadic* relation $B(abc)$ to mean either (aRb and bRc) or (cRb and bRa). Now prove: *If $B(abc)$ holds, then $B(acb)$ does not hold.*

(b) What is the meaning of $B(abc)$ in the three applications of the branch of pure mathematics given in the lecture text?

36.3. Write out the statements of the theorems and definitions of the geometrical application of the branch of pure mathematics of the lecture text.

36.4. Write out the statements of the theorems and definitions of the arithmetical application of the branch of pure mathematics of the lecture text.

36.5. In the branch of pure mathematics of the lecture text, let K be a set of four concentric circles of different radii, and let R mean "is contained within." Show that these meanings yield an interpretation of the postulate set, and write out the statements of the theorems and definitions in the resulting model.

36.6. A dyadic relation which is reflexive, symmetric, and transitive in a set K is called an *equivalence relation* in set K. Show that the following dyadic relations are equivalence relations: "is contemporaneous with" and "is the same age as," applied to a set of people: "is equal to" and "has the same parity as," applied to the set of all natural numbers; "is similar to" and "is congruent to," applied to the set of all triangles of a plane.

36.7. What is wrong with the following argument showing that a relation which is both symmetric and transitive is necessarily reflexive?

By symmetry, aRb implies bRa; and by transitivity, aRb and bRa imply aRa.

36.8. Let S be the set of all ordered pairs of positive integers and define $(a, b) = (c, d)$ if and only if $a + d = b + c$. Show that this definition of equality is an equivalence relation in S.

36.9. Let S be the set of all ordered pairs of integers and define $(a, b) = (c, d)$ if and only if $ad = bc$. Show that this definition of equality is an equivalence relation in S.

36.10. Let S be the set of all nonnegative integers and consider the dyadic relation "a has the same remainder as b when divided by 4." Show that the dyadic relation is an equivalence relation in S.

THE THIRD LEVEL

In building up a branch of pure mathematics, one might think that one can set down an arbitrary collection of symbols for the primitive terms of the discourse and then list an arbitrary collection of statements about these primitive terms for the postulates of the discourse. This is not so. There are certain required and certain desired properties that the system of postulates should possess. This lecture will accordingly be devoted to a brief examination of some of the properties of postulate sets.

There are three distinct levels in axiomatic study. First, there are the concrete axiomatic developments of specific fields of knowledge—these developments are examples of material axiomatics. Second, there are the abstract postulational developments having the specific fields above as models—these developments are examples of formal axiomatics. Third, there is a theory which studies the properties possessed by formal abstract postulational developments. David Hilbert christened this third, and highest, of the three levels, *metamathematics*. The advent of each of the three levels—material axiomatics, formal axiomatics, metamathematics—constitutes a GREAT MOMENT IN MATHEMATICS. The first two have already been considered in our lecture sequence, and it is now time to consider the third one. Though this third level had its beginnings, as we shall see, in the aftermath of the discoveries of a non-Euclidean geometry and a noncommutative algebra in the first half of the nineteenth century, it was not brought into prominence until the publication of Hilbert's *Grundlagen der Geometrie* in 1899, and it was not until about 1920 that it became a well-recognized domain of study.

Of the many properties of postulate systems, we shall here restrict ourselves to the four known as *equivalence, consistency, indepen-*

dence, and *categoricalness*. The first property applies to pairs of postulate systems, and the remaining three apply to individual postulate systems.

Equivalence

Two postulate systems $P^{(1)}$ and $P^{(2)}$ are said to be *equivalent* if each system implies the other, that is, if the primitive terms in each are definable by means of the primitive terms of the other, and if the postulates of each are deducible from the postulates of the other. If two postulate systems are equivalent, then the two abstract studies implied by them are, of course, the same, and it is merely a matter of "saying the same thing in different ways." The idea of equivalent postulate systems arose in ancient times when geometers, dissatisfied with Euclid's parallel postulate, tried to substitute for it a more acceptable equivalent. The modern studies of Euclidean geometry, with their various and quite different postulational bases, clearly illustrate that a postulate system is by no means uniquely determined by the study in question but depends upon which technical terms of the study are chosen as undefined terms and which statements of the study are taken as unproved statements.

There is perhaps no simple criterion for determining which of two equivalent postulate systems is the better; it seems to be largely a matter of personal preference. On grounds of economy of assumption, it might seem that the system which contains fewer primitive terms and fewer postulates is the one to be preferred. On the other hand, on pedagogical grounds, one might prefer that system which more quicky leads to the key theorems of the study.

As a postulate system equivalent to that of the branch of pure mathematics developed in the previous lecture we might employ the same primitive terms along with P1, T1, P3, and P4 as postulates. Since T1 has already been derived from P1, P2, P3, P4, it suffices to show that P2 can be derived from P1, T1, P3, P4. This may be accomplished as follows:

P2. *If aRb, then $a \neq b$.*

Suppose, on the contrary, that we have aRb and $a = b$. Then we also have bRa. But this is impossible by T1. Hence P2 follows by *reductio ad absurdum*.

Consistency

A postulate set is said to be *consistent* if contradictory statements are not implied by the set. This is the most important and most fundamental property of a postulate set; without this property the postulate set is essentially worthless. This is because, if both some statement A and its contradictory statement not-A can be proved, then any statement B whatever can be proved.

The most successful method so far invented for establishing consistency of a postulate set is the method of models. A model of a postulational development, recall, is obtained when we assign meanings to the primitive terms of the system which convert the postulates into true statements about some concept. There are two types of models—concrete models and ideal models. A model is said to be *concrete* if the meanings assigned to the primitive terms are objects and relations chosen from the real world, whereas a model is said to be *ideal* if the meanings assigned to the primitive terms are objects and relations chosen from some other postulate system.

Where a concrete model has been exhibited we feel that we have established the *absolute* consistency of our postulate set, for if contradictory statements are implied by our postulates, then corresponding contradictory statements would hold in our concrete model. But contradictions in the real world we accept as being impossible.

It is not always feasible to try to set up a concrete model of a given postulational development. Thus, if the postulate system contains an infinite number of primitive elements, a concrete model would certainly be impossible, for it appears that the real world does not contain an infinite number of objects. In such instances we attempt to set up an ideal model by assigning to the primitive terms of postulate system A, say, concepts of some other postulate system B, in such a way that the interpretations of the postulates of system A are logical consequences of the postulates of system B. But now our test of consistency of postulate set A can no longer claim to be an absolute test, but only a *relative* test. All we can say is that postulate set A is consistent if postulate set B is consistent and we have reduced the consistency of system A to that of another system B.

Relative consistency is the best we can hope for when we apply the methods of models to many branches of mathematics, for many of the branches of mathematics contain an infinite number of primitive elements. This is true, for example, of Lobachevskian plane geometry. It is possible, however, by setting up a model of Lobachevskian plane geometry within Euclidean plane geometry, to show that the former geometry is consistent if the latter geometry is.

The genealogical model of the simple branch of pure mathematics of the previous lecture, since it is a model within the real world, shows that the postulate set of that branch of pure mathematics is absolutely consistent. Should we, however, replace postulate P4 by

P'4. *K consists of a denumerable infinity of distinct elements.*

We would not be able to establish absolute consistency of the amended postulate set. We can, though, by interpreting the elements of K to be the natural numbers 1, 2, 3, ... , and the relation R to mean "is less than," establish the relative consistency of the amended set. We have obtained a model within the arithmetic of the natural numbers, and therefore the amended system is consistent if the arithmetic of the natural numbers is consistent.

Independence

A postulate of a postulate set is said to be *independent* if it is not a logical consequence of the other postulates of the set, and the entire postulate set is said to be *independent* if each of its postulates is independent.

The most famous consideration in the history of mathematics of the independence of a postulate is that associated with the study of Euclid's parallel postulate. For centuries mathematicians had difficulty in regarding the parallel postulate as independent of Euclid's other postulates (and axioms) and accordingly made repeated attempts to show that it is a consequence of these other assumptions. It was the discovery of, and the ultimate proof of the relative consistency of, Lobachevskian non-Euclidean geometry that finally established the independence of Euclid's parallel postulate. In fact, it is no exaggeration to say that the historical consideration of the independence of Euclid's parallel postulate is responsible for initiating the entire study of properties of postulate sets.

A test for the independence of a postulate of a given postulate set consists in finding an interpretation of the primitive terms of the postulate system which falsifies the concerned postulate but verifies each of the remaining postulates. If we are successful in finding such an interpretation, then the concerned postulate cannot be a logical consequence of the remaining postulates, for if it were a logical consequence of the remaining postulates, then the interpretation which converts all the other postulates into true propostitions would have to convert it also into a true proposition. A test, along these lines, of the independence of an entire set of postulates can apparently be a lengthy business, for if there are n postulates in the set, n separate tests (one for each postulate) will have to be formulated.

Independence of a postulate set is by no means necessary, and a postulate set clearly is not invalidated just because it lacks independence. Generally speaking, a mathematician prefers a postulate set to be independent, for he wants to build his theory on a minimum of assumptions. A postulate set which is not independent is merely redundant in that it contains one or more statements which can appear as theorems instead of as postulates. Sometimes, however, for pedagogical reasons, it may be wise to develop a subject from a postulate set which is not independent—for example, in developing plane geometry in high school from a postulational foundation. An early theorem in the development may be difficult to prove. This theorem can then be stated as one of the postulates. Later, when the students have gained the requisite mathematical maturity and familiarity with the subject, it can be pointed out that the postulate is really not independent and a demonstration of it from the other postulates can be given.

There are some well-known postulate sets which, when first published, unknowingly contained postulates that were not independent. Such was the situation with Hilbert's original set of postulates for Euclidean geometry. This set was later shown to possess two postulates which are implied by the others. The finding of these two dependent postulates in no way invalidated Hilbert's system; in a subsequent amendment these postulates were merely changed to theorems, and their proofs supplied.

Similarly, R. L. Wilder was able to show that R. L. Moore's famous set of eight postulates, which virtually inaugurated modern set-theoretic topology, could be reduced to seven by the elimination of

Moore's sixth postulate. The suspicion that the sixth postulate was not independent arose from the fact that the independence proof for this postulate was found to be at fault, and a subsequent search for a satisfactory proof turned out to be fruitless. Of course Moore's mathematical theory remained intact in spite of Wilder's discovery, but the reduction of an eight-postulate system to an equally effective seven-postulate system has an aesthetic appeal to the mathematician.

Let us show that the postulate set for the simple branch of pure mathematics of the previous lecture is an independent set.

To show the independence of postulate P1, let us interpret K as consisting of two brothers, their father, and their father's father, and interpret the R-relation as "is an ancestor of." This interpretation verifies P2, P3, and P4, but falsifies P1.

To show the independence of postulate P2, we may interpret K as the set of integers 1, 2, 3, 4, and the R-relation as "is not greater than." This interpretation verifies each postulate except P2.

To show the independence of postulate P3, let us interpret K as a set of any four distinct objects and the R-relation as "is not identical with." Now all the postulates except P3 are verified.

Finally, to show the independence of postulate P4, one may interpret K as the set of five integers 1, 2, 3, 4, 5, and the R-relation as "is less than." All the postulates except P4 are verified in this interpretation.

Categoricalness

The property of *categoricalness* is more recondite than the three properties already described, and we must precede its definition by first introducing the notion of *isomorphic interpretations* of a postulate system.

Among the primitive terms of a postulate system P we have a collection of E's, say, which denote elements, perhaps some relations $R_1, R_2, \ldots$ among the elements, and perhaps some operations $O_1, O_2, \ldots$ upon the elements. Accordingly, an interpretation of the postulate system is composed at least in part by element constants (the meanings assigned to the E's), perhaps in part by relation constants (the meanings assigned to the R's), and perhaps in part by operation constants (the meanings assigned to the O's). Now in any given interpretation I of P, let a collection of e's be the element contants (representing the E's), r_1, r_2, ... the relation constants (representing the R's), and o_1, o_2, ... the operation constants (representing the O's); and in any other interpretation I' of P let the element constants be a collection of e'''s, the relation constants be r'_1, r'_2, ..., and the operation constants be o'_1, o'_2, ... If it is possible to set up a one-to-one correspondence between the elements e of I and the elements e' of I' in such a way that, if two or more of the e's are related by some r, the corresponding e'' are related by the corresponding r', and if an o operating on one or more of the e's yields an e, the corresponding o' operating on the corresponding e'''s yields the corresponding e', then we say that the two interpretations I and I' of P are *isomorphic*. This definition is often more briefly stated by saying that two interpretations I and I' of a postulate system P are isomorphic if one can set up a one-to-one correspondence between the elements of I and those of I' in such a way as "to be preserved by the relations and the operations of P." Two isomorphic interpretations of a postulate system P are, except for superficial differences in terminology and notation, identical; they differ from one another no more than does the multiplication table up to 10×10 when correctly written out in words first in English and then in French.

With the notion of isomorphic interpretations of a postulate system established, we are prepared to define categoricalness of a postulate set. A postulate set P, as well as the resulting branch of mathematics, is said to be *categorical* if every two interpretations of P are isomorphic.

Categoricalness of a postulate set is usually established by showing that any interpretation of the postulate set is isomorphic to some given interpretation. This procedure has been applied to Hilbert's postulate set for Euclidean plane geometry; it can be shown that any interpretation of Hilbert's postulates is isomorphic to the algebraic interpretation provided by Descartes' analytic geometry. Lobachevskian plane geometry has also been shown to be a categorical system.

There are advantages and disadvantages in having a system categorical. Perhaps the most desirable feature of a noncategorical postulate set is its wide range of applicability—there is not essentially only one model for the system. For example, the theorems of absolute plane geometry—those theorems common to Euclidean and Lobachevskian plane geometry—may be obtained from Hilbert's postulate set for Euclidean plane geometry with the parallel postulate deleted. This truncated postulate set is, of course, noncategorical, since the Euclidean and Lobachevskian interpretations are nonisomorphic. The same can be said of any postulate set for a group; the set is satisfied by nonisomorphic finite and infinite interpretations. One advantage of categoricalness, on the other hand, is that often theorems of a categorical system may be more easily established by establishing their counterparts in some highly familiar model. For example, there are models of Lobachevskian plane geometry within Euclidean plane geometry. Since Lobachevskian plane geometry is a categorical system, and since we are much more conversant with Euclidean plane geometry than with Lobachevskian plane geometry, there arises the very real possibility of establishing theorems in Lobachevskian plane geometry by establishing their counterparts in the Euclidean model.

It should be quite clear that the postulate set for the simple branch of pure mathematics of the preceding lecture is categorical. If, however, we delete postulate P4 from the postulate set, then the remaining postulate set is noncategorcal. This is readily seen by interpreting R as "is less than" and by taking K to consist in turn of the three integers 1, 2, 3 and the four integers 1, 2, 3, 4. Each of these interpretations verifies postulates P1, P2, and P3, but the interpretations are not isomorphic, since it is impossible to set up a one-to-one correspondence between the three elements of the one interpretation and the four elements of the other interpretation.

The creation of metamathematics—the study of properties of postulate systems—certainly was a GREAT MOMENT IN MATHEMATICS. There are many other properties of postulate systems than those we have discussed. In the next lecture we shall say something about a property called *completeness*. Studies of this property, we shall see, led, in 1931, to a celebrated GREAT MOMENT IN MATHEMATICS lying within the field of metamathematics, that is, to a GREAT MOMENT IN MATHEMATICS within a GREAT MOMENT IN MATHEMATICS.

Exercises

37.1. If p, q, r represent propositions, show that the following set of four statements is inconsistent: (1) If q is true, then r is false. (2) If q is false, then p is true. (3) r is true. (4) p is false.

37.2. Let S be a set of elements and F a dyadic relation satisfying the following postulates:

P1. *If a and b are elements of S and if bFa, then aFb.*

P2. *If a is an element of S, then there is at least one element b of S such that bFa.*

P3. *If a is an element of S, then there is at least one element b of S such that aFb.*

P4. *If a, b, c are elements of S such that bFa and cFb, then cFa.*

P5. *If a and b are elements of S such that bFa, then there is at least one element c of S such that cFa and bFc.*

Show that the statement, "If a is an element of S, then there is at least one element b of S, distinct from a, such that bFa and aFb," is consistent with the above postulates.

(This set of postulates, augmented by the statement above, has been used in relativity theory, where the elements of S are interpreted as *instants* and F as meaning "follows."*)

37.3. Let K be a set of elements and R a triadic relation, and consider the following eight statements: (1) If a, b, c are elements of K, then at least one of the six relations $R(abc)$, $R(bca)$, $R(cab)$, $R(cba)$, $R(bac)$, $R(acb)$ holds. (2) There are at least three elements x, y, z of K such that $R(xyz)$ holds. (3) The two relations $R(abc)$ and $R(acb)$ cannot both hold. (4) If $R(abc)$ holds, then a, b, c are distinct. (5) If $R(abc)$ holds, then $R(bca)$ holds. (6) If $R(xab)$ and $R(ayb)$ hold,

---

* See A. A. Robb, *A Theory of Time and Space*. New York: Cambridge University Press, 1914.

then $R(xay)$ holds. (7) If $R(xab)$ and $R(ayb)$ hold, then $R(xyb)$ holds. (8) If $R(abc)$ holds and x is any other element of K, then either $R(abx)$ or $R(xbc)$ holds. Show that the above eight statements are consistent.

37.4. Show that the following four sets of statements of Exercise 37.3 are equivalent to one another: (1), (3), (4), (5), (6); (1), (3), (4), (5), (7); (1), (3), (4), (5), (8); (2), (3), (4), (5), (8). (Each of these four sets constitutes a postulate set for *cyclic order*.)

37.5. Consider the following postulate set, in which *bee* and *hive* are primitive terms:

P1. *Every hive is a collection of bees.*

P2. *Any two distinct hives have one and only one bee in common.*

P3. *Every bee belongs to two and only two hives.*

P4. *There are exactly four hives.*

Show that this set of postulates is absolutely consistent.

37.6. Show that the statement in quotation marks in Exercise 37.2 is independent of the postulates of the exercise.

37.7. Show that postulates P2, P3, P4 of the postulate set of Exercise 37.5 are independent.

37.8. Show that the postulate set of Exercise 37.5 is categorical.

37.9. Show that the postulate set of the simple branch of pure mathematics of the previous lecture is categorical.

37.10. A dyadic relation R is said to be an *equivalence relation* in a set K of elements $a, b, c, \ldots$ if

P1. *For each a of K, aRa.*

P2. *If aRb, then bRa.*

P3. *If aRb and bRc, then aRc.*

Show that this postulate set is independent.

Further Reading

BLANCHÉ, ROBERT, *Axiomatics*, tr. by G. B. Keene. London: Routledge and Kegan Paul, 1962.

EVES, HOWARD, *A Survey of Geometry*, rev. ed. Boston: Allyn and Bacon, 1972.

MEYER, BURNETT, *An Introduction to Axiomatic Systems*. Boston: Prindle, Weber & Schmidt, 1974.

WILDER, R. L., *Introduction to the Foundations of Mathematics*, 2nd ed. New York: Wiley, 1965.

MATHEMATICS AS A BRANCH OF THEOLOGY

In 1931 there appeared, in the journal *Monatshefte für Mathe-matik und Physik*, a paper entitled, "Über formal unentscheibare Sätze der Principia Mathematica und verwandter Systeme" ("On formally undecidable propositions of Principia Mathematica and related systems"). The author of the paper was a 25-year-old Austrian mathematician and logician named Kurt Gödel, who was, at the time, at the University of Vienna. When the paper appeared, it received only scattered and scant attention, for it concerned itself with a highly specialized area of study that had not yet attracted many researchers and it used a method of proof that was so technically novel as to be incomprehensible to most readers. Within a very few years, however, the paper became widely recognized pro-fessionally as one of the truly epoch-making contributions to the foundations of mathematics and logic. It marked a highly signifi-cant GREAT MOMENT IN MATHEMATICS and led, in 1938, to the appointment of the author as a permanent member of the Institute for Advanced Studies at Princeton and, in 1952, to the rather rare event of a mathematician receiving an honorary degree from a top-flight American university—Harvard.

Gödel's paper revealed certain unforeseen limitations in ax-iomatic procedure. In particular, it accomplished the following: (1) It upset a strong belief that all the important areas of mathematics can be completely axiomatized. (2) It annihilated all hopes of establishing the inner consistency of mathematics along lines that had been envisioned by David Hilbert. (3) It led to a reappraisal, not yet completed, of certain widely held philosophies of mathematics. (4) It introduced into foundational studies a new, powerful, and fertile technique of analysis which has suggested and initiated many new avenues of investigation.

It is the purpose of the present lecture to comment on these four points. Because of the advanced nature of the material and methods involved, the discussion will be brief and will skirt the formidable technicalities of the subject.

1. Consider the hypothetical task of building up a postulate set for Euclidean plane geometry. The first thing we might do is to select our primitive terms. These must constitute a collection of technical terms of the geometry such that all other technical terms of the study can be defined by means of them. The next thing we might do is to begin formulating a growing list of mutually compatible statements about the primitive terms. These will be our postulates. Here a problem arises. When can we cut short our growing list of postulates? We would like our postulate set to be ample enough to imply the "truth" or the "falseness" of any possible statement in Euclidean plane geometry. In other words, we would like to have a sufficient number of postulates so that if S is any statement whatever concerning the primitive and defined terms, then either S or its contradictory statement, not-S, is implied by our postulates. If the postulate set is not sufficiently ample, there will certainly be some statements of the geometry which cannot be reached from our postulates. This condition would exist, for example, if we should choose Hilbert's collection of primitive terms and all of Hilbert's postulates except, say, the parallel postulate. This slightly truncated postulate set could never decide for us whether or not the sum of the angles of a triangle is always equal to 180°, for our truncated postulate set is common to the postulate sets of both Euclidean and Lobachevskian plane geometry. Apparently our list of postulates will be sufficiently ample if a point is reached when it becomes impossible to add to our list any further statement which is both independent of, and consistent with, the postulates so far formulated. If such a point can be reached, then we will have what may be called a complete postulate set for the geometry.

We may now give the following formal definition of completeness. A consistent postulate set is said to be *complete* if it is impossible to add to the set, without extending the collection of primitive terms, another postulate which is both independent of, and consistent with, the given postulates.

Following the development of the concept of formal axiomatics by David Hilbert and others in the early part of the twentieth century, the axiomatic method was vigorously exploited. A number of new and old branches of mathematics were supplied with sets of postulates which were thought to be adequate for the settlement of any question involving the primitive and defined terms of those branches of mathematics; that is, the branches of mathematics were supplied with what were thought to be complete postulate sets. For example, it was thought that the Peano postulate set for the natural number system was complete, or, if not complete, that it could surely be made so by the addition of one or more further postulates. This belief was shattered by Gödel's paper, for in that paper Gödel proved the following theorem:

GÖDEL'S FIRST THEOREM. *For any consistent formal system F which contains the natural number system there are undecidable propositions in F; that is, there are propositions S in F such that neither S nor not-S is provable in F.*

It follows that any postulate set for the natural number system must, if consistent, be incomplete. In other words, no matter what consistent postulate set one might adopt for the natural number system, there will be statements S about the natural numbers such that neither S nor not-S is provable from the postulates. This was a surprising and disappointing discovery.

There are many famous conjectures in number theory that, in spite of extended and strenuous effort, have never been established or refuted. Among these are, for example, the Goldbach conjecture (that every even integer greater than 2 can be expressed as the sum of two primes) and the conjecture known as Fermat's last theorem (that there do not exist positive integers x, y, z such that $x^n + y^n = z^n$ when n is a positive integer greater than 2). It could be that the statements about natural numbers involved in these conjectures are among those propositions which are undecidable in, say, the Peano development of the natural number system, and this could be the reason no one has been able to establish or refute the conjectures. It surely would be some gain if for a given statement about the natural numbers one could at least determine whether or not it is a provable proposition of the system. But here, too, the outlook is melancholy, for in 1936 the American logician Alonzo Church proved the following theorem:

CHURCH'S THEOREM. *For any consistent formal system F containing the natural number system, there exists no effective method of deciding which propositions of F are provable in F.*

2. We saw, in the previous lecture, that consistency of postulate sets is usually achieved by interpretations and models. Such a proof is an indirect process and often merely shifts the question of the consistency of one domain of mathematics to that of another. In other words, a consistency proof by the method of models is often only relative. It is conceivable that absolute consistency may be established by a direct procedure which endeavors to show that by following the rules of deductive inference no two theorems can be arrived at from a given postulate set which will contradict one another. In such a procedure a complete listing of the permissible rules of logic is, of course, necessary. Hilbert attacked the problem of securing the consistency of classical mathematics in such a direct manner. Much as one may prove, by the rules of a game, that certain situations cannot occur within the game, Hilbert hoped to prove, by a suitable set of rules of procedure for obtaining new theorems from previous theorems, that contradictory situations cannot occur in the "game" of mathematics.

Although the direct method of establishing consistency is too complicated to illustrate here, we can clarify the idea of the method by an analogue in chess. Suppose we wish to show that in a game of chess, no matter how many moves might be made, one can never, if one plays in accordance with the rules, arrive at the situation where there are 10 queens of the same color on the board. Here the direct method is applicable, for we can prove from the rules of the game that no move can increase the sum of the number of queens and pawns of the same color. Since this sum is initially 9, it must remain ≤ 9.

Hilbert set about his consistency program with great zeal. But, though he was able to illustrate with certain simple systems what he hoped to do for the more complicated system of all classical mathematics, he was unable to carry out his consistency proof for the latter system. This was shown by Gödel to be inevitable, for in his famous paper he also managed to prove the following theorem:

GÖDEL'S SECOND THEOREM. *For any consistent formal system F which contains the natural number system, the consistency of F cannot be proved in F.*

It follows that among the undecidable problems in F, that of the consistency of F is one. This shattered Hilbert's hopes as originally envisioned, and it now seems that the internal consistency of classical mathematics cannot be attained unless one adopts principles of reasoning of such complexity that the internal consistency of these principles is as open to doubt as that of classical mathematics itself.

3. Gödel's two theorems have shown that no complete axiomatic development of certain important sectors of mathematics is attainable and that no truly impeccable guarantee can be given that certain important sectors of mathematics are free of internal contradiction. These are severe limitations of the axiomatic method, and they point up that the processes of mathematical proof may not, indeed, probably do not, coincide with formal axiomatic procedures. No prior limits can be imposed on the inventiveness of mathematicians in devising new procedures of proof. Apparently human intellectual resources cannot be fully formalized, and new principles of proof await discovery and invention. All this has obvious repercussions in any discussion of a philosophy of mathematics and shows that some currently widely held philosophies of mathematics must be revamped or scrapped.

Gödel's two theorems are certainly among the most remarkable of all metamathematical theorems. As F. De Sua aptly put it,* Gödel's theorems show that "the formal systems known to be adequate for the derivation of mathematics are unsafe in the sense that their consistency cannot be demonstrated by finitary methods formalized within the system, whereas any system known to be safe in this sense

---

*F. De Sua, "Consistency and completeness—a résumé," *American Mathematical Monthly*, 63 (1956), pp. 295–305.

is inadequate." Quoting further from De Sua, we find the following interesting remark: "Suppose we loosely define a *religion* as any discipline whose foundations rest on an element of faith, irrespective of any element of reason which may be present. Quantum mechanics for example would be a religion under this definition. But mathematics would hold the unique position of being the only branch of theology possessing a rigorous demonstration of the fact that it should be so classified."

4. Though the proofs of Gödel's theorems (as supplied by Gödel himself, or later by others, like Barkley Rosser and S. C. Kleene) are too technical to be given here, something should be said of Gödel's remarkable device for "arithmetizing" his treatment. The device has proved to be very powerful and surprisingly applicable, and it somewhat compares, in its area of applicability, with the fertile method of Descartes in geometry.

Gödel's device lies in a method of numbering his primitive symbols, his formulas, and his proofs. In any completely formalized system F, there are, first of all, basic symbols in terms of which propositions and proofs are expressed. For instance, in Gödel's system F, appear (among others) the symbols

$$f, \quad \sim, \quad (, \quad),$$

and a denumerably infinite set of variables in each of a denumerably infinite set of types. Gödel chose to assign the numbers 3, 5, 11, and 13, respectively, to the above symbols, and

$$17^n, 19^n, 23^n, 29^n, \ldots$$

to the variables of type n, where 17, 19, 23, 29, ... is the sequence of consecutive primes after the prime 13. We shall call these numbers, assigned to the basic symbols, *Gödel numbers*.

Now a proposition in F, expressed in the symbols of F, is nothing but a finite ordered sequence of basic symbols of F. Such sequences are called *formulas*, and a unique number is assigned to each formula as follows. Let $n(1), n(2), \ldots, n(s)$ be the Gödel numbers of the symbols of a formula P in the order in which they occur in P, and let $p_1, p_2, \ldots, p_s$ be the first s consecutive primes, starting with

$p_1 = 2$. Then the number assigned to the formula P, which is called the *Gödel number* of P, is the product

$$p_1{}^{n(1)}p_2{}^{n(2)} \cdots p_s{}^{n(s)}.$$

For example, consider the formula (or part of a formula)

$$\sim (x(fy)),$$

where x and y are variables of types 2 and 1, respectively, possessing the Gödel numbers 17^2 and 17. Then the Gödel numbers of the consecutive symbols of the formula are 5, 11, 289, 11, 3, 17, 13, 13, whence the Gödel number of the formula itself is the product

$$2^5 3^{11} 5^{289} 7^{11} 11^3 13^{17} 17^{13} 19^{13}.$$

It follows that for every proposition, or formula, there is a unique Gödel number. Also, given the Gödel number of a formula, the formula itself can be regained. All one has to do is to factor the number of the formula into its unique product of prime factors. Then the number of 2's occurring in the factorization is the Gödel number of the first symbol of the formula, the number of 3's occurring in the factorization is the Gödel number of the second symbol of the formula, the number of 5's occurring in the factorization is the Gödel number of the third symbol in the formula, and so on.

Finally, a proof is merely a finite ordered sequence of formulas, and a device can be used to assign a Gödel number to a *proof* similar to that used to assign a Gödel number to a formula. Suppose, for example, that a proof is made up of the ordered sequence of propositions, or formulas, $P_1, P_2, \ldots, P_t$. Let $f(i)$ be the Gödel number of formula P_i. Then the Gödel number of the proof will be the product

$$p_1{}^{f(1)}p_2{}^{f(2)} \cdots p_t{}^{f(t)},$$

where $p_1, p_2, \ldots, p_t$ is the sequence of the first t consecutive primes, starting with $p_1 = 2$. As before, given the Gödel number of a proof, the steps, or consecutive formulas, of the proof can be regained by factorization.

A very interesting and important feature of Gödel's device is that it permits metamathematical statements—that is, statements about the formal system F—to be "translated" into statements about numbers. For when Gödel numbers are assigned to the formulas of

F, then a statement A about these formulas can be replaced by a statement B about the associated Gödel numbers of the formulas, such that statement B is true if and only if statement A is true. For example, suppose statement A says, "Formula P_1 consists of formula P_2 with some more symbols added at the end." For statement B we may then take, "The Gödel number of P_1 is a factor of the Gödel number of P_2." Now it happens that an important class of these B statements can themselves be expressed as formulas in the system F, with their own Gödel numbers, and the formal system F can, to this extent, "talk about itself." It is this ability to arithmetize the metamathematics of the formal system F that permitted Gödel to prove his two remarkable theorems, which are, of course, metamathematical statements about his system F.

Before closing this lecture, let us give a simple application of the Gödel numbering device. Let us use the device to give another proof of the fact that the set of all rational numbers is denumerable. To this end, we first note that any rational number can be uniquely written in the form $(-1)^n p/q$, where n, p, q are nonnegative integers chosen as small as possible and $q \neq 0$. As examples, we have

$$-18/60 = (-1)^1 3/10 \quad \text{and} \quad 8/12 = (-1)^0 2/3.$$

Now set up the correspondence

$$(-1)^n p/q \leftrightarrow 2^n 3^p 5^q,$$

and call the latter number the *Gödel index* of the rational number $(-1)^n p/q$. As an example, the Gödel index of $-3/4$ is $2^1 3^3 5^4 = 33{,}750$. There is clearly a one-to-one correspondence between the rational numbers and their Gödel indices. The enumeration of the rational numbers is now obtained by listing the rational numbers in the order of magnitude of their Gödel indices.

Exercises

38.1. Can a noncategorical postulate set ever be complete?

38.2. Prove, from the rules of the game, that in a game of tic-tactoe one can never have 6 crosses on the board.

38.3. (a) Find the Gödel number of the formula $(fu(fv))$, where u

and v are variables of types 1 and 2, respectively, possessing Gödel numbers 19 and 19^2.

(b) Find the formula corresponding to the Gödel number

$$2^5 3^{11} 5^3 7^{19} 11^{13}.$$

38.4. Find the Gödel indices of $-12/60$ and $9/36$.

38.5. Find the rational number whose Gödel index is 2250.

Further Reading

KLINE, MORRIS, *Mathematics: The Loss of Certainty*. New York: Oxford University Press, 1980.

NAGEL, ERNEST, and JAMES R. NEWMAN, *Gödel's Proof*. New York: New York University Press, 1958.

WILDER, R. L., *The Foundations of Mathematics*, 2nd ed. New York: Wiley, 1965.

THE DREAM THAT CAME TRUE

About 1812, the eccentric English mathematician and mechanist, Charles Babbage (1792–1871) began to consider the construction of a machine to aid in the calculation of mathematical tables. According to one story, the idea first came to him when the younger Herschel brought in, for checking, some calculations that had been performed for the Astronomical Society. In the course of the tedious checking, Herschel and Babbage found a number of errors, finally causing Babbage to exclaim, "I wish to God these calculations had been executed by steam." "It is quite possible," replied Herschel. From this chance interchange of remarks arose the obsession that was to dominate Babbage for the rest of his life and was to transform him from a lighthearted young man into a cantankerous old one.

In 1822 Babbage pointed out, in a letter to the President of the Royal Society, the advantages to the government of a machine that would calculate the lengthy tables used in navigation and astronomy, and he offered his services in the construction of such a machine for the government's use. His proposal was enthusiastically received, and in 1823 the government agreed to grant funds for the enterprise, which was to take three years.

Babbage threw himself into his project with great zeal and energy. He set to work to make what he called his *difference engine*, capable of employing twenty-six significant figures and of computing and printing successive differences out to the sixth order. But the work did not progress satisfactorily. Babbage was constantly envisioning new and grander ideas about the machine—ideas that frequently led to scrapping all that had already been done and to beginning anew. The result was that after about ten years the governmental aid was withdrawn. Babbage thereupon abandoned his difference engine and commenced work on a much more ambitious machine that he called his *analytical engine*, which was intended to execute completely automatically a whole series of arithmetical calculations assigned to it at the start by an operator. The machine would be able to store intermediate results in its "memory," with a capacity of a thousand numbers of fifty digits each, for future use. It would be able to use auxiliary tabular numbers, like logarithms, which it would possess in its own "library." It would be able to make its own judgments by comparing numbers, and then to act upon these judgments. And all this, which constitutes the essence of a modern computer, was to be carried out purely mechanically, without the use of such modern aids as electricity, vacuum tubes, relay systems, and transistors. Needless to say, the analytical engine, also, was never completed, largely because of the grandness of the project, the lack of needed funds, and the fact that technology had not yet achieved the ability to make the necessary precision tools.

The British government had poured some £17,000 (more than two million dollars in present value) into the construction of the difference engine, and Babbage had contributed a comparable amount. The unfinished machine, and drawings for the complete machine, along with fragments and drawings of the analytical engine, were deposited in the Museum at King's College in London, and then later moved to the South Kensington Science Museum, where they now reside. The part of the difference engine on exhibit is still in good working order and not long ago was taken apart, thoroughly cleaned, and reassembled so that an exact copy could be made for the museum of the International Business Machines (IBM) Corporation.

Though the Babbage projects failed, they did provide the inspiration for the remarkable giant mechanical and electronic high-speed computers that have come into existence in recent years. Babbage, who was a hundred years ahead of his time, had enunciated the principles on which all modern computing machines are based. When the British magazine *Nature* published an article in 1946 discussing one of America's first large electronic calculators (the Harvard relay computer, *Mark I*), it entitled the article, "Babbage's Dream Comes True."

The early history of successful computing devices can be quickly told. Beyond the computational aid given to man by nature in the form of his ten fingers (and still used by children in school classrooms today), the highly efficient and inexpensive abacus of ancient origin (still widely used in many parts of the world), and the calculating rods designed by John Napier in 1617 (employed in the schools today as a pedagogical device), the invention of the first calculating machine is attributed to Blaise Pascal, who, in 1642, devised an adding machine to assist his father in the auditing of the government accounts at Rouen. This instrument was able to handle numbers not exceeding six digits. It contained a sequence of engaging dials, each marked from 0 to 9, so designed that when one dial in the sequence turned from 9 to 0 the preceding dial of the sequence automatically turned one unit. Thus the "carrying" process of addition was mechanically accomplished. Pascal manufactured over fifty machines, some of which are still preserved in the Conservatoire des Arts et Métiers at Paris.

Later in the century, Leibniz (1671) in Germany and Sir Morland (1673) in England invented machines that multiply. Similar attempts were made by a number of others, but most of these machines proved to be slow and impractical. In 1820, Thomas de Colmar, although not familiar with Leibniz's work, transformed a Leibniz type of machine into one that could perform subtractions and divisions. This machine proved to be the prototype of almost all commercial machines built before 1875, and of many developed since that time. In 1875, the American Frank Stephen Baldwin was granted a patent for the first practical calculating machine that could perform the four fundamental operations of arithmetic without any resetting of the machine. In 1878, Willgodt Theophile Odhner, a Swede, was granted a United States patent on a machine very similar in design to that of Baldwin. The various makes of electrically operated desk calculators of recent times, such as the Friden, Marchant, and Monroe, have essentially the same basic construction as the Baldwin machine. Remarkable and useful as these desk calculators became, they are, however, vastly inferior in speed, scope, and applicability to the great electronic computers of today. Unlike the desk calculators, which owe their ultimate parentage to the 1642 adding machine of Blaise Pascal, these later computers owe their parentage to Babbage's analytical engine of the 1830's.

One of the first direct descendants of Babbage's analytical engine is the great *Automatic Sequence Controlled Calculator* (the ASCC), known as *Mark I*, constructed under the guidance of Professor Howard Aiken of Harvard University, and opened to the public in 1944, as a joint enterprise by the university and IBM under contract for the Navy Department. The machine is 51 feet long, 8 feet high, with two panels 6 feet long, and weighs 5 tons. An improved second model of the ASCC, known as *Mark II*, was made for use, beginning in 1948, at the Naval Proving Ground, Dahlgren, Virginia. Another descendant of Babbage's effort is the *Electronic Numerical Integrator and Computer* (the ENIAC), a multipurpose electronic computer completed in 1945 at the University of Pennsylvania under contract with the Ballistic Research Laboratory of the Army Proving Ground, Aberdeen, Maryland. This was the first digital computer controlled by vacuum tubes. The machine requires a 30 ft. by 50 ft. room, contains 19,000 vacuum tubes, and weighs 30 tons. It may now be viewed in the Smithsonian Institution in Washington, D.C. These amazing high-speed computing machines, along with similar projects, like the *Selective Sequence Electronic Calculator* (SSEC) of IBM, the *Electronic Discrete Variable Calculator* (EDVAC) of the University of Pennsylvania, the MANIAC of the Institute for Advanced Study at Princeton, the *Universal Automatic Computer* (UNIVAC) of the Bureau of Standards, and the various *differential analyzers*, presaged machines of even more fantastic accomplishment. It was in the 1940's, largely as the work of the American mathematician John von Neumann (1903-1957), that the idea developed of storing a computer's program in the machine's memory. UNIVAC I, which was built in 1951, became the first of a variety of computers that were mass-produced in the 1950's; the manufacture of computers had become an industry.

Every few years, a new generation of machines eclipses in speed, reliability, and memory those of the preceding generation. The following table of comparisons of calculations of π performed on electronic computers shows the rapid increase in computational speed that has taken place.

Most of the early high-speed computers were designed to solve military problems, but today machines are also being designed for banks, businesses, government, transportation systems, engineer-

Author	Machine	Date	Decimal Places	Time	
Reitwiesner	ENIAC	1949	2037	70	hours
Nicholson and Jeenel	NORC	1954	3089	13	minutes
Felton	Pegasus	1958	10000	33	hours
Genuys	IBM 704	1958	10000	100	minutes
Genuys	IBM 704	1959	16167	4.3	hours
Shanks and Wrench	IBM 7090	1961	100265	8.7	hours

ing, and many other purposes. From luxury tools they have become vital and necessary instruments in present-day development. Because of this, numerical analysis has received tremendous stimulus in recent times and has become a subject of ever-growing importance. It is becoming not uncommon for secondary schools to offer introductory courses in computer science and computer programming and to have either a modest computer of their own or a tie-up with some large computer located at a nearby college or university. Babbage's dream certainly has come true.

Unfortunately, there is a developing feeling, not only among the general public but also among young students of mathematics, that from now on any mathematical problem will be resolved by a sufficiently sophisticated electronic machine and that essentially all mathematical activity today is computer oriented. Teachers of mathematics must combat this disease of *computeritis*—by repeatedly pointing out that the machines are merely extraordinarily fast and efficient symbol manipulators and are invaluable in mathematics only where extensive symbol manipulation, as in lengthy calculations and exhaustive enumeration and testing of patterns and cases, is involved.

Nonetheless, in their area of applicability, the machines have scored some spectacular mathematical successes. For example, the recent accomplishments, reported in LECTURE 11, concerning amicable, perfect, and prime numbers, would have been virtually impossible without the assistance of a computer. These machines have proved themselves valuable not only in parts of number theory but in many other mathematical studies, such as group theory, finite geometries, graph theory, matrix work, numerical solution of differential equations, and recreational mathematics, to mention a few.

For example, in the field of recreational mathematics, in 1958 Dana S. Scott instructed the MANIAC digital computer to search out all solutions of the problem of putting together all twelve pentominoes* to form an 8 × 8 square with a 2 × 2 hole in the center. After operating for about 3.5 hours, the machine produced a complete list of 65 distinct solutions, wherein no solution can be obtained from another by reflections and rotations. Similarly, the enumeration and construction of all 880 distinct normal magic squares† of order 4 is easily achieved with a computer, and it is not difficult to program a machine for the corresponding problem for magic squares of order 5. Though this latter program has been made, no one has yet carried it through on a computer. It has been estimated that the number of distinct normal magic squares of order 5 is about 15,000,000. The teasing problem of dissecting a geometrical square into subsquares, no two of which are equal, has also been successfully approached by a computer, yielding some results that have so far evaded noncomputer procedures.

Let us look briefly at two deeper and more serious mathematical problems that have been attacked by computers. We have seen, above, some remarkable machine calculations of π out to large numbers of decimal places. For five years the computation of π to slightly over 100,000 decimal places made by Daniel Shanks and J. W. Wrench, Jr., on an IBM 7090, remained unbeaten. Then, on February 22, 1966, M. Jean Guilloud and his co-workers at the Commissariat à l'Énergie Atomique in Paris attained an approximation of π extending to 250,000 decimal places, using a STRETCH computer. Exactly one year later, the workers found π to 500,000 places using a CDC 6600, and in 1974 the same group found π to 1,000,000 decimal places using a CDC 7600.

There is more to the calculation of π to a large number of decimal places than just the challenge involved. One reason is to secure

---

*A *pentomino* is a planar arrangement of five unit squares joined along their edges.

†A *normal magic square of order n* is any arrangement of the first n^2 positive integers into an $n \times n$ square array such that the numbers along any row, column, or main diagonal have the same sum. Two magic squares are *distinct* if neither can be obtained from the other by reflections and rotations.

statistical information concerning the "normalcy" of π. A real number is said to be *simply normal* if in its decimal expansion all ten digits occur with equal frequency, and it is said to be *normal* if all blocks of digits of the same length occur with equal frequency. It is not known if π (or even $\sqrt{2}$ for that matter) is normal or even simply normal. The calculations of π, starting with that on the ENIAC in 1949, were performed to secure statistical information on the matter. From counts of these extensive expansions of π, it would seem that π is perhaps normal. An earlier erroneous 707-place calculation of π made by William Shanks of England in 1873 seemed to indicate that π was not even simply normal.

The matter of the normalcy or nonnormalcy of π will never, of course, be resolved by computers. We have here an example of a theoretical problem which requires profound mathematical talent and which cannot be solved by computation alone. The existence of such problems ought to furnish at least a partial antidote to the rampant disease of computeritis.

The Dutch mathematician L. E. J. Brouwer (1882–1966), seeking, for logical and philosophical purposes, a mathematical question so difficult that its answer in the following ten or twenty years would be very unlikely, finally hit upon: "In the decimal expansion of π, is there a place where a thousand consecutive digits are all zero?" If π is normal, as is suspected, then a block of 1000 zeros will occur in its decimal expansion not only once but infinitely often and with an average frequency of 1 in 10^{1000}. It follows that much more is required to prove that π is normal than just to answer Brouwer's question, and this latter appears in itself to be a considerable task.

The elaborate calculations of π have another use in addition to furnishing statistical evidence concerning the normalcy or nonnormalcy of π. Every new automatic computing machine, before it can be adopted for day-to-day use, must be tested for proper functioning, and coders and programmers must be trained to work with the new machine. Checking into an already-found extensive computation of π is frequently chosen as an excellent way of carrying out this required testing and training.

Having considered the part played by computing machines in approximating π to a large number of decimal places in order to obtain statistical information concerning the normalcy or nonnormalcy of

π, we will conclude our lecture with a brief account of what is perhaps the most impressive mathematical victory achieved by the machines—the resolution, in the summer of 1976, of the famous "four-color conjecture."

About 1850 Francis Guthrie,* when a graduate student at University College in London, noticed that four colors are sufficient to distinguish the counties on a map of England. Somewhat later he showed his younger brother, Frederick Guthrie,† then a student of Augustus De Morgan,‡ an unsatisfactory argument (now lost) that four colors will suffice to color any map on a plane or a sphere, where two countries sharing a common boundary have different colors. This conjecture has since become known as the *four-color problem*.

Frederick Guthrie communicated the problem to his teacher De Morgan, who, in turn, in a letter dated October 23, 1852, communicated it to Sir William Rowan Hamilton and stated that he was unable to supply a proof. Hamilton exhibited no interest in the matter, and so for a time the problem lay dormant. Then, on June 13, 1878, at a meeting of the London Mathematical Society, Arthur Cayley announced he had been unable to obtain a proof of the conjecture. In the first volume of the *Proceedings of the Royal Geographical Society* (1879), Cayley again stated the problem.

Shortly after Cayley's announcement, A. B. Kempe, a British barrister-at-law, published, in 1879, a "proof" of the conjecture in the *American Journal of Mathematics*. A simplified version of the "proof" was published later in the same year in the *Transactions of the London Mathematical Society*, and again, in the following year, 1880, in *Nature*. In 1880, in the *Proceedings of the Royal Society of Edinburgh*, appeared a reduction of the problem, by P. G. Tait, Professor of Mathematics at the University of Edinburgh, to an unproved property about a closed graph. In 1890, in the London

---

*Francis Guthrie (1831–1899), a former student of Augustus De Morgan, became a Professor of Mathematics at South African University in Cape Town.

†Frederick Guthrie (1833–1866) later became Professor of Chemistry and Physics at the newly created School of Science, South Kensington.

‡Augustus De Morgan (1806–1871) was a professor at University College in London and founder of the London Mathematical Society.

Quarterly Journal of Mathematics, P. J. Heawood* pointed out a flaw, which had evaded detection for 11 years, in Kempe's reasoning. For close to 100 years this flaw remained uncircumvented, and for that long period of time the four-color problem stood as one of the most celebrated unverified conjectures in mathematics. Heawood's work was far from being entirely destructive, for during his lifetime he contributed a great deal to various aspects of the four-color problem, and he was the first to prove that five colors will certainly suffice for coloring any map on a plane or a sphere.

In 1920, Philip Franklin showed that all maps on a plane or a sphere of 25 or fewer countries can be colored with four colors. In 1926, C. N. Reynolds raised this number to 27, then Franklin, in 1936, raised it to 31. C. E. Winn, in 1943, raised the number to 35, and Oystein Ore and Joel Stemple, in 1968, raised it to 40.

Many mathematicians worked on the four-color problem, discovering various reduction schemes that limit the type of map that needs to be considered and finding a number of challenging alternative equivalent forms of the conjecture, but, in spite of many purported "proofs," the problem itself remained refractory. The growing complexity of investigations led to suggestions of programming some aspect or other of the problem for electronic high-speed computers.

Then, in the summer of 1976, Kenneth Appel and Wolfgang Haken of the University of Illinois established the conjecture by an immensely intricate computer-based analysis. Their proof contains several hundred pages of complex detail and subsumes over 1000 hours of computer calculation. The proof involves an examination of 1,936 reducible configurations, each requiring a search of up to half a million logical options to verify reducibility. This last phase of the work occupied six months and was finally completed in June 1976. Final checking, which itself had to be done by a computer, took almost the entire month of July, and the results were communicated to the *Bulletin of the American Mathematical Society* on July 26, 1976.

---

*P. J. Heawood (1861-1955) was an English mathematician who probably spent a longer span of time on the four-color problem than anyone else.

The Appel-Haken solution is unquestionably an astounding accomplishment, but a solution based upon computerized analyses of close to 2000 cases with a total of a billion logical options is very far indeed from elegant mathematics. Certainly on at least an equal footing with a solution to a problem is the elegance of the solution itself. This is probably why, when the result above was personally presented by Haken to an audience of several hundred mathematicians at the University of Toronto in August 1976, the presentation was rewarded with little more than a mildly polite applause.

Of course it is possible that, in time, someone may find a proof of the four-color problem independent of any computer analysis—a proof elegant and concise enough to be verifiable by an unassisted human mind. Nevertheless, one begins to wonder if mathematics perhaps contains problems beyond this realization, that is, contains problems of such complexity that they are quite beyond an unassisted human mind and therefore *must* be approached via some computer. There is every reason to believe that such problems exist. From this point of view, the four-color problem is perhaps more valuable to mathematics than to cartography, in that it may help clarify possible limitations of purely human solutions to problems.

Each of three earlier aids to calculation—the abacus, the Hindu-Arabic numeral system, and logarithms—earned inclusion in our lectures as GREAT MOMENTS IN MATHEMATICS. As a further great aid, the modern electronic high-speed computer, prophesied a hundred years earlier by Charles Babbage, also merits inclusion. And among the astonishing accomplishments of the computers, the resolution of the four-color conjecture in the summer of 1976 must itself be considered a GREAT MOMENT IN MATHEMATICS.

Very useful to students, businessmen, and engineers are the pocket-size calculators now available for less than $50 and becoming less expensive and more sophisticated each year. These little machines, handling numbers of about 8 digits, possessing a memory, and able instantly to perform any arithmetic operation and, in some cases, trigonometric calculations, owe their remarkable compactness to the miniature transistors, chips, etc., that have replaced the electronic vacuum tubes of earlier and larger machines. They have made slide rules obsolete instruments.

Exercises

39.1. Babbage's difference engine was designed to construct mathematical tables by utilizing successive differences. To illustrate the method, let us construct a table of the squares 1^2, 2^2, 3^2, 4^2, ... of the succcessive positive integers. We set up three columns, A, B, and C, the first two of which are work columns. Column C is the list of sought squares; column B is the list of differences of column C; column A is the list of differences of column B. Note that (in our illustration) all entries in column A are 2's.

$$
\begin{array}{ccc}
A & B & C \\
  &   & 1 \\
  & 3\rightarrow & \overline{} \\
2\rightarrow & \overline{} & \overline{4} \\
  & 5\rightarrow & \\
2\rightarrow & \overline{} & \overline{9} \\
  & 7\rightarrow & \\
  &   & \overline{16}
\end{array}
$$

We need to specify only the three initial entries, 2, 3, and 1, of these columns, along with the following fixed pattern of procedure based upon the single operation of addition. Complete column A by filling in 2's as far down as we wish. Now complete column B by adding in the entries from column A as indicated by the arrows and addition bars. Finally, in the same way, complete column C by adding in the entries of column B. Babbage's difference engine was capable of automatically performing the operations of this pattern, and thus of constructing the sought table of squares.

(a) Set up the initial entries for calculating a table of the values of $2n^2 - n + 1$ for the successive positive integers n.

(b) Set up the initial entries for calculating a table of values of n^3 for the successive positive integers n.

39.2. (a) How many distinct triominoes are there?

(b) How many distinct tetrominoes are there?

(c) Construct patterns of the twelve distinct pentominoes.

(d) There are many challenging puzzles involving the twelve pentominoes. For example, all twelve can be assembled into a 6×10, or a 5 × 12, or a 4 × 15, or a 3 × 20 rectangle. Half of them can be assembled into a 5 × 6 rectangle and the other half into another 5 × 6 rectangle. We leave these assemblages to any interested reader.

39.3. (a) The sum of the numbers in any column, row, or main diagonal of a magic square is called the *magic sum* of the square. What is the magic sum of a normal magic square of order n?

(b) Show that the central number of a normal magic square of the third order must be 5.

(c) Show that in a normal magic square of the third order 1 can never occur in a corner position.

39.4. (a) Is 2/7 simply normal?

(b) Can a rational number ever be simply normal?

39.5. (a) Construct a planar map of four countries that requires four colors to color it.

(b) Prove that a map formed by n circles in a plane can be colored with two colors.

Further Reading

GOLDSTINE, H. H., *The Computer from Pascal to Von Neumann*. Princeton, N.J.: Princeton University Press, 1972.

MORRISON, PHILIP and EMILY, *Charles Babbage and His Calculating Engines* (*selected writings of Charles Babbage and others*). New York: Dover, 1961.

STEEN, L. A., editor, *Mathematics Today: Twelve Informal Essays*. New York: Springer-Verlag, 1978.

APOLOGY AND REGRETS

In our written lecture sequence of forty-three GREAT MOMENTS IN MATHEMATICS many fine candidates for inclusion had to be passed by. The original oral lecture sequence contained a choice of sixty-some GREAT MOMENTS, but even then there still were many important omissions. We deeply regret the necessity for these exclusions and sincerely apologize for the brevity and incompleteness of the written accounts of those items that were finally chosen. By way of slight amends, and before bidding the patient reader of these lectures adieu, it seems fitting at least to mention, in chronological order and with a word or two, some of the omitted GREAT MOMENTS IN MATHEMATICS, each of which constitutes a mathematically thrilling story. Some of these omitted GREAT MOMENTS are concerned with the growing content of mathematics over the ages, while others are more concerned with the changing nature of mathematics over the ages. Most of those of the former category subsequently found more extended treatment in Howard Eves, *An Introduction to the History of Mathematics,* and most of those of the latter category subsequently found a more extended treatment in Howard Eves and C. V. Newsom, *An Introduction to the Foundations and Fundamental Concepts of Mathematics.* These books, both published by Holt, Rinehart and Winston (the latest edition of the first in 1976 and that of the second in 1965), may be consulted by an interested reader.

*   *   *   *   *

1. *Plimpton 322* (sometime from 1900 to 1600 B.C.). Plimpton 322, meaning the item with catalogue number 322 in the G. A. Plimpton archaeological collection of Columbia University, is certainly one of the most interesting of the ancient Babylonian mathe-

matical cuneiform tablets yet analyzed. Written in Old Babylonian script, it dates sometime from 1900 to 1600 B.C., and was first described by Otto Neugebauer and A. J. Sachs in 1945. Apparently one of a set of three tablets (the other two have not yet been found), Plimpton 322 shows the astonishing development of mathematics in the Mesopotamian crescent in those very early times. The exciting analysis of the tablet reveals a knowledge of the Pythagorean relation, a procedure for finding all primitive Pythagorean triples, and a method of constructing a table of trigonometric secants.

2. *Zeno's paradoxes* (ca. 450 B.C.). Should we assume that a magnitude is infinitely divisible or rather that it is made up of a very large number of small indivisible atomic parts? Some of the logical difficulties encountered in either assumption were strikingly brought out in the fifth century B.C. by four clever paradoxes devised by the Eleatic philosopher Zeno. These paradoxes have had a profound influence on mathematics, particularly on the subsequent development of the calculus.

3. *Aristotle's systematization of deductive logic* (ca. 340 B.C.). A mathematical system is the resultant of two components, a set of postulates and a logic (see Figure 19). That is, the theorems of a mathematical system result from the interplay of an initial set of statements, called the postulates, and another initial set of statements, called the logic or rules of procedure for obtaining consequences of assumptions. The first to give a systematic consideration of logic was Aristotle (384–322 B.C.) of Stageira, one-time tutor of

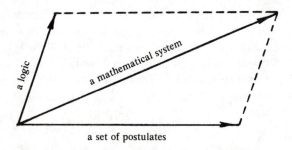

a set of postulates

FIG. 19

Alexander the Great and successor to Plato. Further significant study of logic did not occur until modern times.

4. *The "Conic Sections" of Apollonius* (ca. 225 B.C.). No work more overwhelmingly exhibits the geometrical skill of the ancient Greeks than the comprehensive *Conic Sections* of the great geometer Apollonius of Perga. In this work Apollonius supplied the names *ellipse, parabola,* and *hyperbola,* and established a vast collection of properties of these curves, the bulk of which he derived from the geometrical equivalents of Cartesian equations of the curves. This is an outstanding example of a mathematical study that was pursued for its own intrinsic interest and beauty and with no thought of practical application. More than eighteen hundred years later, Kepler found here the curves and properties that he needed for the formulation of his three famous laws of planetary motion.

5. *Al-Khowârizmî's contributions* (ca. 820). During the wise and colorful reign (809–833) of caliph al-Mâmûn, the scholar Mohammed ibn Mûsâ al-Khowârizmî (Mohammed, son of Moses of Khowarezm) wrote a treatise on algebra and a book on arithmetic. These works are perhaps the most influential of all the early Arabian contributions to mathematics. The latter work, when translated into Latin in the twelfth century, played a cardinal role in the introduction and dissemination of the Hindu-Arabic numeral system in Western Europe.

6. *The trigonometry of Regiomontanus* (ca. 1464). Johann Müller (1436–1476), more commonly known as Regiomontanus, after the Latinized form of his birthplace of Königsberg ("king's mountain"), was the ablest and most influential mathematician of the fifteenth century. His five-book treatise *De triangulis omnimodis,* written about 1464 but not published until 1533, is his greatest work and constitutes the first systematic European exposition of plane and spherical trigonometry considered independently of astronomy. Though the only trigonometric functions employed are the sine and cosine, and though the algebra involved is rhetorical, any high school student with an interest in trigonometry will find pleasure and challenge in some of the propositions of Regiomontanus' work.

7. *Simon Stevin and decimal fractions* (1585). Of the five great inventions to assist calculation—the abacus, the Hindu-Arabic

numeral system, decimal fractions, logarithms, and the modern computer—all but that of decimal fractions have already appeared in our work as GREAT MOMENTS IN MATHEMATICS. It is quite fitting that the invention of decimal fractions also be so considered. Though this invention cannot be assigned to any single individual, one of the earliest and ablest expositors of decimal fractions, and the most influential mathematician of the Low Countries in the sixteenth century, was Simon Stevin (1548–1620), who became quartermaster of the Dutch army and director of many public works. Simon Stevin was a very interesting man from a number of points of view. His work on decimal fractions was published in 1585, in a Flemish edition titled *La Thiende* and a French edition titled *La Disme*.

8. *Thomas Harriot, founder of the English school of algebraists* (1631). Thomas Harriot (1560–1621), of particular interest to Americans because in 1585 he was sent by Sir Walter Raleigh to the New World to survey and map what was then called Virginia but is now North Carolina, is the acknowledged founder of the English school of algebraists. His great work in this field, the *Artis analyticae praxis*, was not published until ten years after his death and deals largely with the theory of equations. It was this work that served as the prototype for subsequent texts on the theory of equations. Earlier in the present century, virtually every college and university in the United States offered a course in the subject; today's mathematical curricula find almost no place for it.

9. *Desargues and the birth of projective geometry* (1639). In 1639 there appeared, in Paris, a remarkably original but eccentrically written and little-heeded treatise on the conic sections, written by Gérard Desargues (1593–ca. 1662), an engineer, architect, and onetime French army officer. Some two centuries later, the work was resurrected by the French geometer and historian Michel Chasles, and since then it has been recognized as the first classic in the early development of synthetic projective geometry.

10. *Academies, societies, and periodicals* (1662, 1666). The great increase in scientific and mathematical activity in the seventeenth century, and the inadequacy of disseminating findings by written correspondence, led to the formation of discussion circles and, later, to full-fledged societies and academies, with regular times of meet-

ing devoted to the presentation and discussion of scholarly papers. Many of these groups, in time, began the publication of periodicals for even wider circulation of discoveries. The British Royal Society was founded in London in 1662 and the French Academy of Sciences in Paris in 1666. Though there had been a few small earlier assemblages, these two groups mark the serious start of the academies and societies. Prior to 1700, there were only 17 periodicals containing mathematical material. The subsequent history and development of mathematical societies and periodicals constitutes a highly interesting and exciting story.

11. *Johann Bernoulli and the calculus of variations* (1696). In 1696–97 Johann Bernoulli (1667–1748) proposed and solved the problem of the *brachystochrone*, namely, the determination of the curve of quickest descent of a weighted particle moving between two given points, not on the same vertical line, in a gravitational field. This is an example of a higher type of maximization-minimization problem than those encountered in the differential calculus. In the differential calculus we seek a *point* which yields some extreme value; here we are seeking a *curve* which yields an extreme value. The field of mathematics devoted to extrema of this second kind is called the calculus of variations.

12. *Poncelet and the golden period of projective geometry* (1822). Jean Victor Poncelet (1788–1867), a French army officer, was taken prisoner of war during Napoleon's retreat from Moscow. While held in Russia for two years, with no books at hand, Poncelet planned his great *Traité des propriétés projectives des figures,* which, after his release and return to France, he published in Paris in 1822. This work gave tremendous impetus to the study of projective geometry and inaugurated the so-called "golden period" in the history of the subject.

13. *The principle of duality* (1826). The beautiful principle of duality of plane projective geometry, which pairs the propositions of the subject, was first explicitly stated by Joseph-Diez Gergonne (1771–1859) in 1826, though it was approached by the works of Poncelet and others of the first quarter of the nineteenth century. Once the principle of duality is in some way established, the proof of one proposition of a dual pair automatically carries with it a proof of the

other. It has since been shown that other areas of mathematics possess principles of duality, for example, the study of trigonometric equations, the geometry of spherical triangles, solid projective geometry, Boolean algebra, the calculus of propositions, and the theory of partially ordered sets. Julius Plücker (1801–1868) established the principle of duality analytically in 1829, when he developed a plane analytic geometry in which lines have coordinates and points have linear equations.

14. *Transcendental numbers* (1851, 1873, 1882). Many students of mathematics never encounter proofs of certain great classical results, like the fundamental theorem of algebra, the transcendence of e and π, the impossibility of trisecting a general angle, duplicating a cube, and squaring a circle with Euclidean tools, and so on. Quite elementary proofs of many of these results can be given. For example, in connection with transcendental numbers, a sophomore college mathematics student can easily follow proofs of the following interesting results:

(1) $t = \sum\limits_{k=1}^{\infty} a_k / 10^{k!} = 0.a_1 a_2 000 a_3 00000000000000000 a_4 000 \ldots$,

where the a_k's are arbitrary digits from 1 to 9, is a transcendental number. [First shown by Joseph Liouville (1809–1882) in 1851.]

(2) e is transcendental. [First established by Charles Hermite (1822–1901) in 1873.]

(3) π is transcendental. [First established by Ferdinand Lindemann (1852–1939) in 1882.]

15. *Hilbert's Paris problems* (1900). David Hilbert (1862–1943) was invited to deliver a major address in Paris, in the summer of 1900, at the second International Congress of Mathematicians. After much thought, he decided upon a lecture that would attempt to look into the future of mathematics. To this end he chose twenty-three highly fruitful unsolved problems whose solutions, he prophesied, would engage mathematicians during the coming century and would mark out important lines in the future development of mathematics. A discussion of these problems and subsequent attempts to solve them constitute great pages in the history of recent mathematics.

16. *The Lebesgue integral* (1902). The fundamental theorem of calculus states the inverse relation between differentiation and integration—a relation that was shown to hold so long as the functions being integrated are continuous. The preservation of the fundamental theorem for more general functions was one of the first fruits of a new theory of integration introduced by Henri Lebesgue (1875–1941) in his 1902 doctoral thesis and expanded by him in his 1904 book. Based upon a novel partitioning of the range of the function, rather than its domain, many functions that are not integrable in the older Riemann sense become integrable in the newer Lebesgue sense, resulting in a great enlargement of the class of integrable functions. Other and further generalizations of the integral were subsequently given by Denjoy, Haar, and Stieltjes.

17. *Mathematical logic* (1910–1913). It would be a hopeless task to discuss modern considerations of logic by the use of only ordinary language. A symbolic language has become necessary to achieve an exact and unambiguous treatment of the subject. Because of the presence of such symbolism, the resulting treatment is known as *symbolic,* or *mathematical, logic.* Though Leibniz was perhaps the first seriously to consider the desirability of a symbolic logic, and the idea was furthered by such men as George Boole, Augustus De Morgan, Charles Sanders Peirce, Ernst Schröder, and Gottlieb Frege, the monumental *Principia mathematica* (1910–1913) of Alfred North Whitehead (1861–1947) and Bertrand Russell (1872–1970) marks an unusually cardinal and influential achievement. A remarkable feature of this work is that it approaches the calculus of propositions by the postulational method. A small selection is made from the set of all tautologies (laws of logic) to serve as the postulates of the development, and then several formal rules are given in accordance with which all other tautologies can be obtained from the selected few. The rules play the same role in the development of the propositional calculus that logical inference plays in the usual development of a branch of mathematics. Of course, logical inference, in the conventional sense, cannot be used here, for it is this logical inference that now constitutes the object of study.

18. *Multiple-valued logics* (1921). Since, in classical logic and the symbolic logic of *Principia mathematica,* a proposition can assume

228 GREAT MOMENTS IN MATHEMATICS (AFTER 1650)

any one of two possible truth values, namely *true* or *false,* these logics are called *two-valued logics.* In 1921, in a short two-page paper, J. Lukasiewicz considered a three-valued logic, or a logic in which a proposition *p* may possess any one of three possible truth values. Very shortly after, E. L. Post considered *m*-valued logics, in which a proposition *p* may possess any one of *m* possible truth values, where *m* is any integer greater than 1. Another study of *m*-valued logics was given in 1930 by Lukasiewicz and A. Tarski. Then, in 1932, the *m*-valued truth systems were extended by H. Reichenbach to an infinite-valued logic, in which a proposition *p* may assume any one of infinitely many possible truth values. These new logics are not barren of application—they have been applied to the mathematical theory of probability, to the quantum theory of modern physics, to the establishment of the independence of the postulates of familiar two-valued logic, and elsewhere. The denial by Lukasiewicz and Post of Aristotle's famous law of excluded middle (that limits a proposition to just two truth values) reminds one of the denial by Lobachevsky and Bolyai of Euclid's axiom of parallels, and the denial by Hamilton and Cayley of the axiom that claims multiplication is commutative.

19. *Bourbaki* (1939). Since 1939 a comprehensive set of volumes on mathematics, starting with the most general basic principles and proceeding into various specialized areas, has been appearing in France under the alleged authorship of a Nicolas Bourbaki. Bourbaki must be ranked as one of the most influential mathematicians of our century; his works are much read and quoted. He has enthusiastic supporters and scathing critics. And, most curious of all, he does not exist. For Nicolas Bourbaki is a collective pseudonym employed by an informal group of mathematicians. Though the members of the organization have taken an oath of anonymity, their names are largely an open secret to mathematicians. It is believed that among the original members were C. Chevalley, J. Delsarte, J. Dieudonné, and A. Weil. The membership has varied over the years, sometimes numbering as many as twenty mathematicians.

20. *Nonstandard analysis* (1960). The calculus, as developed in the seventeenth and eighteenth centuries, involved infinitely small quantities called *infinitesimals,* which, in the words of Johann Ber-

noulli, are so small that "if a quantity is increased or decreased by an infinitesimal, then that quantity is neither increased nor decreased." This seemingly contradictory situation offended mathematicians' feelings of rigor, and accordingly infinitesimals were banned from mathematics and their use replaced by ϵ, δ procedures and a theory of limits. In 1960 Abraham Robinson succeeded in putting infinitesimals and their application to the calculus on a rigorous basis. This accomplishment, which greatly simplifies the calculus, has both pedagogical and theoretical values, and may prove to be one of the major mathematical achievements of the present century. It can be used to justify much of the "loose" reasoning of Euler and other early writers, and it can be employed in teaching a first course in the calculus.

HINTS FOR THE SOLUTION OF SOME
OF THE EXERCISES

21.1. (a) 15 : 1.

21.1. (b) 21 : 11.

21.2. (a) See any College Algebra text.

21.2. (b) Use 21.2 (a).

21.2. (c) Use 21.2 (a).

21.3. (a) By the binomial theorem, the sought coefficient is

$$n(n - 1) \cdots (n - r + 1)/r!.$$

Now multiply numerator and denominator by $(n - r)!$ and use 21.2 (a).

21.3. (b) In the binomial expansion of $(a + b)^n$ set $a = b = 1$.

21.4. (a) This follows from the definition of the arithmetic triangle as given in the lecture text.

21.4. (b) Use successive applications of 21.4 (a).

21.4. (c) Use mathematical induction, 21.4 (a), and 21.2 (a).

21.4. (d) Use 21.4 (c).

21.4. (e) Use 21.4 (a).

21.4. (f) Use 21.4 (e).

21.4. (g) Use 21.4 (c).

22.1. (a) Use the fact that the sum of the focal radii of a point on an ellipse is constant.

22.1. (b) Use the fact that the difference of the focal radii of a point on a hyperbola is constant.

22.2. Here we have

$$(x - x_2)^2 + y^2 = (1 - x_2)^2 + 4.$$

230

The elimination of y gives

$$(x - x_2)^2 + 4x = (1 - x_2)^2 + 4,$$

or

$$x^2 + 2x(2 - x_2) + (2x_2 - 5) = 0.$$

The condition that this quadratic equation have two equal roots is that its discriminant vanish, that is, that

$$(2 - x_2)^2 - (2x_2 - 5) = 0$$

or $x_2 = 3$. The required tangent may now be drawn.

22.3. Consider the tangent at the point (x_1, y_1) as the limiting position of the secant line through (x_1, y_1) and a neighboring point

$$(x_1 + \Delta x, y_1 + \Delta y)$$

of the curve as $\Delta x \to 0$.

22.4. (a) One finds $t = -y^2/x$. Slope $= y/t = -x/y = -3/4$.

22.4. (b) One finds slope $= a/e = -x/y = -3/4$.

22.4. (c) One finds slope $= \dot{y}/\dot{x} = -x/y = -3/4$.

22.4. (d) One has $y = \sqrt{25 - x^2}$, whence

$$dy/dx = -2x/2\sqrt{25 - x^2} = -3/4.$$

22.5. Consult any introductory calculus text.

22.6. Employ mathematical induction.

22.7. ds/dt is the time rate of change of distance and d^2s/dt^2 is the time rate of change of ds/dt.

22.8. At a turning-point maximum or minimum, the tangent line is horizontal, and therefore has zero slope. The condition is not sufficient, since for $y = x^3$ we have $dy/dx = 0$ at $(0,0)$, but there is neither a maximum nor a minimum at this point.

23.1. The operation of taking cube roots.

23.3. $\ln x + C$.

23.4. $-\cos x + C$ and $\sin x + C$.

23.5. $\ln(x^2 - 3x + 5) + C.$

23.6.

$$d(\tan x)/dx = \frac{d}{dx}\left(\frac{\sin x}{\cos x}\right)$$

$$= \left[\cos x \frac{d}{dx}(\sin x) - \sin x \frac{d}{dx}(\cos x)\right]/\cos^2 x$$

$$= \frac{\cos^2 x + \sin^2 x}{\cos^2 x} = \frac{1}{\cos^2 x} = \sec^2 x.$$

Therefore

$$\int \sec^2 x\, dx = \tan x + C.$$

23.7. $d(\tan x - x)/dx = \sec^2 x - 1 = \tan^2 x.$ Therefore

$$\int \tan^2 x\, dx = \tan x - x + C.$$

23.8. 81/4.

23.9. 4.17.

23.10. 2.

24.1. (a) $S = n[2a + (n-1)d]/2.$

24.1. (b) $S = a(1 - r^n)/(1 - r).$

24.1. (c) In the formula for S in Exercise 24.1 (b), let $n \to \infty$.

24.2. (a) Let x be the largest share and d the common difference in the arithmetic progression. Then we find $5x - 10d = 100$ and $11x - 46d = 0.$

24.2. (b) Apply Exercise 24.1 (b).

24.2. (c) Apply Exercise 24.1 (c).

24.3. (a) Recall that $\sum_{i=1}^{n} i = n(n+1)/2.$

24.5. (a) $e^x = 1 + x + x^2/2! + x^3/3! + \cdots$

$\sin x = x - x^3/3! + x^5/5! - x^7/7! + \cdots$

$\cos x = 1 - x^2/2! + x^4/4! - x^6/6! + \cdots.$

24.5. (c) Take $x = \pi$ in Exercise 24.5 (b).

24.6. 1.6487.

24.7. 0.71934.

24.9. 0.764.

24.10. 0.747.

25.2. Merely for convenience, so that the formula (2) of the lecture text for a_n will hold for $n = 0$ as well as for $n = 1, 2, \cdots$, as is seen by integrating

$$f(x) = \frac{a_0}{2} + \sum_{n=1}^{\infty} (a_n \cos nx + b_n \sin nx)$$

termwise between $-\pi$ and $+\pi$ and then using Exercise 25.1.

25.3. One finds

$$a_0 = \frac{1}{\pi} \int_{-\pi}^{\pi} f(x)\, dx = \frac{1}{\pi} \int_{-\pi}^{0} 2\, dx + \frac{1}{\pi} \int_{0}^{\pi} dx = 2 + 1 = 3,$$

$$a_n = \frac{1}{\pi} \int_{-\pi}^{\pi} f(x) \cos nx\, dx = \frac{1}{\pi} \int_{-\pi}^{0} 2 \cos nx\, dx + \frac{1}{\pi} \int_{0}^{\pi} \cos nx\, dx$$

$$= \left[\frac{2}{\pi n} \sin nx \right]_{-\pi}^{0} = \left[\frac{1}{\pi n} \sin nx \right]_{0}^{\pi} = 0, \, n = 1, 2, \ldots,$$

$$b_n = \frac{1}{\pi} \int_{-\pi}^{\pi} f(x) \sin nx\, dx = \frac{1}{\pi} \int_{-\pi}^{0} 2 \sin nx\, dx + \frac{1}{\pi} \int_{0}^{\pi} \sin nx\, dx$$

$$= \left[-\frac{2}{\pi n} \cos nx \right]_{-\pi}^{0} + \left[-\frac{1}{\pi n} \cos nx \right]_{0}^{\pi}$$

$$= -\frac{2}{\pi n} + \frac{2}{\pi n} \cos n\pi - \frac{1}{\pi n} \cos n\pi + \frac{1}{\pi n}$$

$$= \frac{1}{\pi n} (\cos n\pi - 1), \, n = 1, 2, \ldots .$$

Thus all the a's vanish except a_0, and all the b's with even subscripts vanish (since the cosine of an even multiple of π is unity). We find

$$b_1 = -\frac{2}{\pi}, b_3 = -\frac{2}{3\pi}, b_5 = -\frac{2}{5\pi}, \ldots .$$

25.4. $x/2|x| = -1$ for $x < 0$ and $+1$ for $x > 0$.

25.5. The positive roots of $\cos x = 0$ are $x = \pi/2, 3\pi/2, 5\pi/2,$ .... Now use the sum for $\pi^2/8$ found in Example 2 of the lecture text.

25.7. In the Fourier series set $x = \pi/2$.

26.1. To deduce Euclid's fifth postulate, let AB and CD be cut by a transversal in S and T, respectively, and suppose $\angle BST + \angle DTS < 180°$. Through S draw straight line QSR, making $\angle RST + \angle DTS = 180°$. By I 28 (if a transversal of two straight lines makes a pair of interior angles on the same side of the transversal equal to two right angles, then the two lines are parallel), QSR is parallel to CD. Therefore, by Playfair's postulate, AB is not parallel to CD and AB and CD must intersect. By I 17 (any two angles of a triangle are together less than two right angles), AB and CD cannot meet on that side of ST not containing angles BST and DTS.

26.3. Let ABC be any right triangle and draw the perpendicular CD to the hypotenuse AB. Then triangles ABC and ACD have two angles of one equal to two angles of the other, whence the third angles must also be equal. That is, $\angle B = \angle ACD$. Similarly, $\angle A = \angle BCD$. It follows that $\angle A + \angle B + \angle C = 180°$. Now, if ABC is not a right triangle, divide it into two right triangles by an altitude.

26.4. Try the same experiment and reasoning on a spherical triangle, using a great circle arc in place of the straightedge.

26.5. The "proof" assumes that if two triangles have two angles of one equal to two angles of the other, then the third angles are also equal. This, in turn, assumes the existence of noncongruent similar triangles.

HINTS FOR THE SOLUTION OF SOME OF THE EXERCISES

26.6. (a) Draw the diagonals of the isosceles birectangle.

26.6. (b) Draw the diagonals of each "half" of the isosceles birectangle and apply Exercise 26.6 (a).

26.6. (c) Draw a perpendicular, from the vertex of the triangle, upon the line joining the midpoints of the two sides of the triangle.

26.6. (d) Draw the diagonals, radiating from the intersection of the two concerned lines, of the two lower "quarters" of the isosceles birectangle and apply Exercise 26.6 (b).

26.7. (d) In Figure 20 we have

$$
\begin{aligned}
\text{defect } ABC &= 180° - (\alpha_1 + \alpha_2 + \beta_2 + \gamma_1) \\
&= 360° - (\alpha_1 + \beta_1 + \gamma_1 + \alpha_2 + \beta_2 + \gamma_2) \\
&= [180° - (\alpha_1 + \beta_1 + \gamma_1)] + [180° - (\alpha_2 + \beta_2 + \gamma_2)] \\
&= \text{defect } ADC + \text{defect } ABD.
\end{aligned}
$$

26.8. See Exercise 26.6 (b).

26.9. Angle $A < 60°$.

26.10. Consult any standard text on elementary solid geometry.

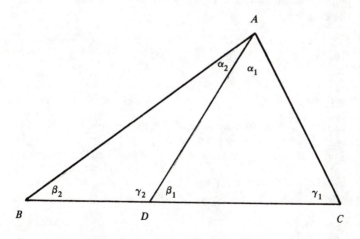

Fɪɢ. 20

27.1. (d) We have

"distance" PQ + "distance" QR

$$= \log \frac{(QS)(PT)}{(PS)(QT)} + \log \frac{(RS)(QT)}{(QS)(RT)}$$

$$= \log \frac{(QS)(PT)(RS)(QT)}{(PS)(QT)(QS)(RT)} = \log \frac{(RS)(PT)}{(PS)(RT)}$$

$$= \text{"distance" } PR.$$

27.1. (e) We have

$$\lim_{Q \to T} \text{"distance" } PQ = \lim_{Q \to T} \log \frac{(QS)(PT)}{(PS)(QT)} = \infty.$$

27.3. (a) One might interpret "abba" as "committee" and "dabba" as "committee member," and assume that there are just two committees and that no committee member serves on more than one committee.

27.3. (b) One might interpret a "dabba" as any one of the three letters a, b, c, and an "abba" as any one of the three pairs ab, bc, ca of these letters.

27.4. See Wolfe, H. E., *Introduction to Non-Euclidean Geometry*, pp. 174–176.

28.1. (a) No.

28.1. (b) Yes.

28.4. (a) Replacing a by z' in the first equality, and by z in the second equality, we find $z' \oplus z = z'$, $z \oplus z' = z$. Therefore $z' = z' \oplus z = z \oplus z' = z$.

28.4. (b) Add each member of the equality $a \oplus b = a \oplus c$ to $\bar{a}$, and then use the associative law for $\oplus$.

28.4. (c) Show that $x = \bar{a} \oplus b$ is a solution; then show, by 28.4 (b), that if there are two solutions, x and y, we must have $x = y$.

28.4. (d) By the distributive law,

$$(a \otimes a) \oplus (a \otimes z) = a \otimes (a \oplus z) = a \otimes a = (a \otimes a) \oplus z.$$

Therefore, by 28.4 (b), $a \otimes z = z$.

28.4. (e) Consider the triple sum $(a \otimes b) \oplus (a \otimes \bar{b}) \oplus (\bar{a} \otimes \bar{b})$. Combine the first two summands, and use the distributive law; then combine the last two summands, and use the distributive law.

28.4. (f) Since $a \otimes b = z = a \otimes z$, we have, if $a \neq z$, by the cancellation law for the operation $\otimes$, $b = z$.

28.4. (g) Show that $x = a^{-1} \otimes b$ is a solution; then show, by the cancellation law for the operation $\otimes$, that if there are two solutions, x and y, we must have $x = y$.

28.4. (h) Show that $(a \otimes b) \oplus (a \otimes \bar{b}) = z$.

28.5. (e) Since $a \ominus b$, $a \oplus \bar{b}$ is positive. Since $b \ominus c$, $b \oplus \bar{c}$ is positive. Therefore, by P12, $(a \oplus \bar{b}) \oplus (b \oplus \bar{c})$ is positive. But $(a \oplus \bar{b}) \oplus (b \oplus \bar{c}) = a \oplus \bar{c}$. Hence $a \oplus \bar{c}$ is positive, and $a \ominus c$.

28.5. (g) Since $a \ominus b$, $a \oplus \bar{b}$ is positive. Since c is given positive, $(a \oplus \bar{b}) \otimes c$ is positive, by P12. But $(a \oplus \bar{b}) \otimes c = (a \otimes c) \oplus (\bar{b} \otimes c)$, by 28.4 (h). Therefore $(a \otimes c) \ominus (b \otimes c)$.

28.5. (h) This follows from 28.4 (e) and 28.4 (h).

28.5. (j) Use 28.5 (i).

28.7. We have $a = a \otimes u$ (by P7) $= a \otimes (c \otimes c^{-1})$ (by P10) $= (a \otimes c) \otimes c^{-1}$ (by P4) $= (b \otimes c) \otimes c^{-1}$ (by substitution) $= b \otimes (c \otimes c^{-1})$ (by P4) $= b \otimes u$ (by P10) $= b$ (by P7).

28.8. Examples (b), (c), (k) are fields.

29.1. (a) $*$ is neither commutative nor associative; $|$ is both commutative and associative; the distributive law holds.

29.1. (b) None of the laws hold.

29.1. (c) $|$ is associative and the distributive law holds.

29.1. (d) Only the two commutative laws hold.

29.3. (b) $(1, 0, -2, 3) (1, 1, 2, -2) = (11, -1, 3, 3)$; and $(1, 1, 2, -2) (1, 0, -2, 3) = (11, 3, -3, -1)$.

29.4. (b)

$$AB = \begin{bmatrix} -4 & -5 \\ -8 & 11 \end{bmatrix}, \quad BA = \begin{bmatrix} 4 & 8 \\ 12 & 3 \end{bmatrix}, \quad A^2 = \begin{bmatrix} -8 & -9 \\ 12 & -11 \end{bmatrix}.$$

29.4. (e) There are no divisors of zero; the left cancellation law for multiplication.

29.5. (a) Show that

$$\begin{bmatrix} 0 & 1 \\ 0 & 0 \end{bmatrix} = \begin{bmatrix} a & b \\ c & d \end{bmatrix}^2$$

implies: (1) $b(a + d) = 1$, (2) $c(a + d) = 0$, (3) $a^2 + bc = 0$, (4) $cb + d^2 = 0$. From (1) it follows that $a + d \neq 0$. Therefore, from (2), $c = 0$. Hence, from (3) and (2), $a = d = 0$. This contradicts that $a + d \neq 0$.

30.2. (a) Yes. (b) No. (c) No. (d) No; G1 does not hold. (e) Yes.

30.5. No.

30.10. By G2' we are guaranteed the existence of an element i such that, for a given element b, $b * i = b$. Now let a be any element of G. By G2' there exists an element c such that $a = c * b$. Then

$$a * i = (c * b) * i = c * (b * i) = c * b = a,$$

and G2 is established. Finally, by G2', there exists for each element a of G an element a^{-1} of G such that $a * a^{-1} = i$, and G3 is established.

30.11. No. Let G be the set of real linear functions of the form $a = a_1x + a_2, a_1 \neq 0$, and let $a * b$ mean $a (db/dx)$, where db/dx is the derivative of b with respect to x. Then x is a right identity element and x/a_1 is a left inverse of a.

31.1. Pairs (a), (b), (e) are commutative.

31.2. Find the product $(B^{-1}A^{-1})(AB)$.

31.3. (b) Rotation of $180°$ about the origin; reflection in a line.

31.4. (a) $C'B' = (ACA^{-1})(ABA^{-1}) = AC(A^{-1}A)BA^{-1} = A(CB)A^{-1}$.

31.4. (b) $(ABA^{-1})^{-1} = AB^{-1}A^{-1}$ (by Exercise 31.2).

31.4. (c) $ABA^{-1} = (AB)A^{-1} = (BA)A^{-1} = B(AA^{-1}) = B$, etc.

31.4. (d) Apply Exercises 31.4 (a) and 31.4 (b).

31.5. Show that the product of any two such transformations, and the inverse of any such transformation, are such transformations.

31.6. (a) Show that the product of any two such transformations, and the inverse of any such transformation, are such transformations.

31.6. (b) Take $k = 1$ in the representation of a planar Lorentz transformation.

31.6. (c) Show that the product of any two lorotations about (c, d), and the inverse of any lorotation about (c, d), are lorotations about (c, d).

31.6. (d) Show that the line $rx + sy + t = 0$ is carried onto the line

$$(r/k)x + sky + rc(1 - 1/k) + sd(1 - k) + t = 0.$$

31.6. (e) See Exercise 31.6 (d).

31.6. (f) Show that the circle $(x - c)^2 + (y - d)^2 = r^2$ is carried onto the ellipse

$$\frac{(x - c)^2}{k^2 r^2} + \frac{(y - d)^2}{r^2/k^2} = 1.$$

31.6. (g) The area of a counterclockwise triangle with vertices (x_1, y_1), (x_2, y_2), (x_3, y_3) is given by

$$(1/2) \begin{vmatrix} x_1 & y_1 & 1 \\ x_2 & y_2 & 1 \\ x_3 & y_3 & 1 \end{vmatrix}.$$

31.6. (h) Eliminate k from the equations

$$x = ka + c(1 - k), \qquad y = (b/k) + d(1 - 1/k).$$

31.7. (a) The square of the distance between the two points.

31.7. (b) The tangent of the angle from the first line to the second line.

31.7. (c) The distance from the point to the line.

31.7. (d) The power of the point with respect to the circle.

31.8. (a) Show, for example, that $RH = D'$ and $HR = D$.

31.8. (b) I, R', H, V, D, D' are self-inverse; R and R'' are inverses.

32.1. (a) $(2n - \frac{1}{2})\pi \le x \le (2n + \frac{1}{2})\pi$, n any integer.

32.1. (b) One real solution if $a \ne 0$; no real solution if $a = 0$ and $b \ne 0$; infinitely many real solutions if $a = 0$ and $b = 0$.

32.2. (a) $\sqrt{a}\,\sqrt{b} = \sqrt{ab}$ if and only if not both a and b are negative.

32.2. (b) $\sqrt{x - y} = i\sqrt{y - x}$ if and only if $x - y$ is nonpositive.

32.3. $-5/2$, not $-2/3$.

32.4. The theorem should read: "If two fractions are equal and have equal *nonzero* numerators, then they also have equal denominators."

32.5. If the two members of an inequality are multiplied by the same negative number, the direction of the inequality changes; log $(1/2) < 0$.

32.6. The integral is improper, since the integrand is discontinuous at $x = 0$.

32.7. Examine for endpoint maxima and minima.

32.8. The terms of an infinite series may be bracketed to suit one's pleasure if and only if the series is absolutely convergent.

32.9. $(a, b) = (c, d)$ if and only if $a + d = b + c$,
$(a, b) + (c, d) = (a + c, b + d)$,
$(a, b) (c, d) = (ac + bd, ad + bc)$.

32.10. $(a, b) = (c, d)$ if and only if $ad = bc$,
$(a, b) + (c, d) = (ad + bc, bd)$,
$(a, b) (c, d) = (ac, bd)$.

33.1. Let M_1 be the midpoint of AB, M_2 the midpoint of M_1B, M_3 the midpoint of M_2B, etc. Denote by E the set of all points on $[AB]$ with the exception of points A, B, M_1, M_2, M_3, ... Then we have

$[AB]$ is composed of E, A, B, M_1 M_2, M_3, ... ,
$(AB]$ is composed of E, B, M_1, M_2, M_3, ... ,
$[AB)$ is composed of E, A, M_1, M_2, M_3, ... ,
(AB) is composed of E, M_1, M_2, M_3, ...

It is now apparent how we may put the points of any one of the four segments in one-to-one correspondence with the points of any other one of the four segments.

33.2. Employ Figure 21.

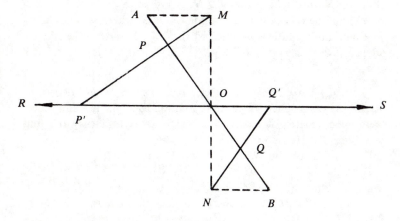

Fig. 21

33.3. Employ Figure 22.

33.4. (c) It is only the verification of the triangle inequality that presents any difficulty. Denote $d(y, z)$, $d(x, z)$, $d(x, y)$ by u, b, c, respectively. Then we have

$$b/(1 + b) = 1/(1/b + 1) \leq 1/[1/(c + a) + 1]$$
$$= (c + a)/(1 + c + a) = c/(1 + c + a)$$
$$+ a/(1 + c + a) \leq c/(1 + c) + a/(1 + a).$$

Notice that for this metric all *distances* are less than 1.

33.5. (b) It is only the verification of the triangle inequality that presents any difficulty. Toward this verification note that, for any three points (x_1, y_1), (x_2, y_2), (x_3, y_3), we have

$$|x_3 - x_1| \leq |x_2 - x_1| + |x_3 - x_2|,$$
$$|y_3 - y_1| \leq |y_2 - y_1| + |y_3 - y_2|.$$

33.7. (b) Postulate H3 is not satisfied for points B and C.

33.8. Since x is a limit point of S, any neighborhood N_x of x contains a point y_1 of S, where $y_1 \neq x$. By Postulate H4, there then exist neighborhoods N_{y_1} of y_1 and N'_x of x such that $N'_x \cap N_{y_1} = \emptyset$. Again, by Postulate H2, there exists a neighborhood N''_x of x such that $N''_x \subset (N_x \cap N'_x)$. It follows that $y_1 \notin N''_x$. But, since x is a limit point of S, N''_x, and hence N_x, contains a point y_2 of S, where $y_2 \neq x$ and $\neq y_1$. Continuing in this way, we find that N_x contains an infinite sequence of distinct points $y_1, y_2, y_3, \ldots$ of S, and the desired result is established.

33.10. (b) Let x be any point of a metric space M and denote by $S(x, r)$ the open sphere of center x and radius r. We now show that these open spheres, if regarded as neighborhoods, satisfy the four postulates of a Hausdorff space.

We omit the obvious verification of H1 and H2.

To verify H3, let $d(x, y) < r$, and set $R = r - d(x, y) > 0$. The triangle inequality states that $d(x, y') \leq d(x, y) + d(y, y') = (r - R) + d(y, y') < r$, if $d(y, y') < R$. That is, $S(y, R) \subset S(x, r)$.

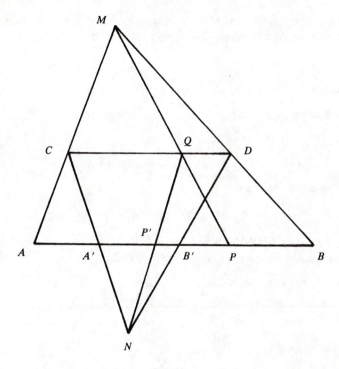

FIG. 22

To verify H4, let x be distinct from y and set $r = d(x, y) > 0$. Then it is easy to show that the two open spheres $S(x, r/3)$ and $S(y, r/3)$ have no point in common.

33.11. From M'1 and M'2 we shall deduce the following two theorems.

THEOREM 1. $d(x, y) = d(y, x)$.

By Postulate M'2 we have

$$d(x, y) \leq d(y, z) + d(z, x)$$

and,

$$d(y, x) \leq d(x, w) + d(w, y).$$

Taking $z = x$ in the first of these inequalities, and $w = y$ in the second one, we find (recalling Postulate M'1)

$$d(x, y) \le d(y, x), \quad d(y, x) \le d(x, y).$$

It now follows that $d(x, y) = d(y, x)$.

THEOREM 2. $d(x, y) \ge 0$.

By M'2, for any point z,

$$d(x, z) \le d(z, y) + d(y, x).$$

Taking $z = x$ we have

$$d(x, x) \le d(x, y) + d(y, x).$$

But $d(x, x) = 0$ (by M'1), and $d(y, x) = d(x, y)$ (by Theorem 1). It follows that $0 \le 2d(x, y)$, whence $d(x, y) \ge 0$.

33.12. (g) See Figure 23.

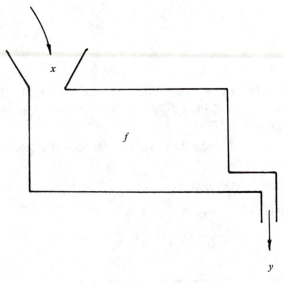

FIG. 23

34.1. (b) Use the idea employed in the proof of Theorem 1 of the lecture text.

34.1. (c) Use an indirect argument along with 34.1 (a) and Theorem 1 of the lecture text.

34.1. (d) Use an indirect argument along with 34.1 (a) and Theorem 2 of the lecture text.

34.5. (a) A circle through three points with rational coordinates has a center with rational coordinates.

34.5. (b) See Problem E832, *American Mathematical Monthly*, 56 (1949), p. 407.

34.5. (c) No, for there are c points on a straight line and on a circle, and there are only d rational numbers and d algebraic numbers.

34.5. (d) Take a number axis on the given straight line. In each interval choose a point with rational coordinate. These points are distinct, and therefore in one-to-one correspondence with the intervals, and they constitute an infinite subset of the denumerable set of all rational numbers.

34.5. (e) Place upon a Cartesian plane and choose within each circle a point having rational coordinates, etc.

34.6. (a) For the rational number a/b, consider the polynomial $bx - a$.

34.6. (b) Consider the polynomial $x^2 - 2$.

34.6. (c) Algebraic, since it is a zero of the polynomial $x^2 + 1$.

34.6. (d) If $\pi/2$ is not transcendental, then it is algebraic and is a zero of some polynomial $f(x)$. Then π would be a zero of the polynomial $f(x/2)$.

34.6. (e) If $\pi + 1$ is not transcendental, then it is algebraic and is a zero of some polynomial $f(x)$. Then π would be a zero of the polynomial $f(x - 1)$.

34.6. (f) If a is transcendental, then so also are a/n and $a + n$, where n is a natural number.

34.6. (g) Multiply the polynomial through by the lowest common denominator of the a_k's.

34.7. The one-to-one correspondence $n \leftrightarrow n + 1$ shows that $d = d + 1$. The one-to-one correspondences $n \leftrightarrow 2n$ and $n \leftrightarrow 2n - 1$ show that $d + d = d$, or $2d = d$.

34.8. With the finite sequence $\{n_1, n_2, \ldots, n_r\}$ associate the natural number

$$n = 2^{n_1} 3^{n_2} \cdots p_r^{n_r},$$

where p_r is the rth prime number. Since factorization of a natural number into powers of primes is unique, there is a one-to-one correspondence between the set of all finite sequences of nonnegative integers and the set of all natural numbers.

34.9. Assume S is denumerable. Then we can arrange the members of S in a sequence $\{f_1(x), f_2(x), \ldots, f_n(x), \ldots\}$. Consider the array

$$
\begin{array}{cccc}
f_1(1) & f_1(2) & f_1(3) & \cdots \\
f_2(1) & f_2(2) & f_2(3) & \cdots \\
f_3(1) & f_3(2) & f_3(3) & \cdots \\
\cdot & \cdot & \cdot & \cdots
\end{array}
$$

Now form the function $f(x)$ such that $f(n) = f_n(n) + 1$. Then $f(x)$ belongs to S. But $f(x)$ cannot be in the given denumeration, for it differs from $f_1(x)$ in the value taken by $x = 1$, from $f_2(x)$ in the value taken by $x = 2$, and so on.

35.4. (a) 48 miles per hour.

35.4. (b) 2.4 days.

35.4. (c) 15 for 68 cents.

35.4. (d) 67.5 cents.

35.4. (e) The second one.

35.4. (f) At the end of 59 seconds.

35.4. (g) A fantastic salary.

35.4. (h) 11 seconds.

35.4. (i) Five cents.

35.4. (j) Neither; the amounts are equal.

35.4. (k) The final pile will be over 17,000,000 miles high.

35.4. (l) No.

35.4. (m) One third.

35.4. (n) Yes.

35.5. The primitive terms are *any* things that satisfy the postulates.

35.6. No, for all four parts.

36.1. Since aGc, there exists m in K such that aFm and mFc, and since bGc, there exists n in K such that bFn and nFc. Since mFc and nFc, by T6 $n = m$. Since aFm and bFm, by T6, $a = b$.

36.2. (a) Employ *reductio ad absurdum* and argue by cases.

36.7. See R. A. Rosenbaum, "Remark on equivalence relations," *American Mathematical Monthly,* 62 (1955), p. 650.

37.1. Suppose q is true. Then, by (1), r is false. But this contradicts (3). Suppose q is false. Then, by (2), p is true. But this contradicts (4).

37.2. Interpret the elements of S as a set of all rectangular Cartesian frames of reference which are parallel to one another but with no axis of one frame coincident with an axis of another frame, and let bFa mean that the origin of frame b is in the first quadrant of frame a. Or, interpret the elements of S as the set of all ordered pairs (m, n) of real numbers, and let $(u, v) F (m, n)$ mean $u > m$ and $v > n$.

37.3. Interpret K as the set of all points on a given circle, and let $R(abc)$ mean "points a, b, c lie in clockwise order."

37.5. Interpret the bees as six people A, B, C, D, E, F, and the four hives as the four committees (A, B, C), (A, D, E), (B, F, E), and (C, F, D).

37.6. Let S be the set of points of a horizontal straight line, and let F mean "is to the right of."

37.7. To show independence of P2, interpret the bees and the hives as four trees and four rows of trees forming the vertices and sides of a square. To show the independence of P3, interpret the bees as four trees located at the vertices and the foot of an altitude of an equilateral triangle, and the hives as the four rows of trees along the sides and the altitude of the triangle. To show independence of P4, interpret the bees and the hives as three trees and three rows of trees forming the vertices and sides of a triangle.

37.8. The following three theorems can be deduced from the postulates of Exercise 37.5: (1) There are exactly six bees. (2) There are exactly three bees in each hive. (3) For each bee there is exactly one other bee not in the same hive with it. Taking these theorems into account, let us designate the four hives by I, II, III, IV and the three bees in hive I by A, B, C. Let hive II have bee A, and only bee A, in common with hive I. Then we may designate the bees in hive II by A, D, E. Now let hive III be that hive which has bee B in common with hive I. Then we may designate the bees in hive III by either B, D, F or B, E, F. In the first case, hive IV must contain bees C, E, F, and, in the second case, hive IV must contain bees C, D, F. There are, then, the following two ways of designating the hives and the bees:

I	A, B, C			I	A, B, C
II	A, D, E		or	II	A, D, E
III	B, D, F			III	B, E, F
IV	C, E, F			IV	C, D, F

But the second designation can be changed into the first by interchanging the two labels D and E. It follows, then, that any interpretation of the postulate set can be labeled as in the first designation, and the desired isomorphism is established.

37.9. Show that there is essentially only one way of labeling the four elements of K.

37.10. See W. T. Guy, Jr., "On equivalence relations," *American Mathematical Monthly* 62 (1955), pp. 179–180.

38.1. No.

38.2. Since noughts and crosses are put on the board alternately, and since the total possible number of noughts and crosses is 9, there can never be more than 5 crosses on the board at any one time.

38.3. (a) $2^{11}\ 3^3\ 5^{19}\ 7^{11}\ 11^3\ 13^{261}\ 17^{13}\ 19^{13}$.

38.3. (b) $\sim (fu)$.

38.4. 18,750.

38.5. $-2/3$.

39.1. (a) The initial column entries are 4, 5, 2.

39.1. (b) The initial column entries are 6, 12, 7, 1.

39.2. (a) Two.

39.2. (b) Four.

39.3. (a) $n(n^2 + 1)/2$.

39.3. (b) Denote the numbers in the magic square by letters and then add together the letters of the middle row, the middle column, and the two main diagonals.

39.3. (c) Use 39.3 (b) and an indirect argument.

39.4. (a) No.

39.4. (b) Yes, for example $r = 0.\overline{1234567890}$.

39.5. (b) Employ mathematical induction.

INDEX

Ab initio process, 25
Abstract spaces, 150, 151
  Hausdorff, 152
  metric (*see* Metric space)
  taxicab, 151
  topological, 153
Academies, societies, periodicals, 224, 225
Acta eruditorum, 19, 19*n*
Aiken, H., 212
Alexander the Great, 223
Algebraic number, 164
Algebraic structure, 88, 91, 92, 110, 115
Al-Khowârizmî, 223
American Journal of Mathematics, 216
Analysis situs of the plane, 126
Analytical engine, 210, 211, 212
Apollonius, 11, 223
  Conic Sections, 11, 223
Appel, K., 217, 218
Applied mathematics
  branch of, 175, 185, 186
  to-date, 175
Archimedes, 42, 43, 49, 227
  Quadrature of the Parabola, 42
Area under a curve, 28, 29
Aristotle, 159, 222, 228
  law of excluded middle, 228
  systematization of deductive logic, 222
Arithmetic algebra, 91

Arithmetic progression, 41
Arithmetical triangle, 6*ff*, 9, 10
Arithmetization of analysis, 138
Ars conjectandi (Bernoulli), 8
Artis analyticae praxis (Harriot), 224
ASCC (Automatic Sequence Controlled Calculator)
  Mark I, 212
  Mark II, 213
Associative law of addition, 89, 99, 101, 105
Associative law of multiplication, 89, 99, 101, 105
Astragalus bone, 1
Ausdehnungslehre (Grassmann), 103
Axioms (or postulates) of a discourse, 173, 174, 177

Babbage, C., 209, 210, 212, 213, 218, 219
  analytical engine, 210, 211, 212
  difference engine, 209, 219
Baldwin, F. S., 211
  calculating machines, 211
Ballistic Research Laboratory, 212
Barrow, I., 16, 17, 34, 36
  differential triangle, 17
  differentiation, 16
  Lectiones opticae et geometricae, 16, 34
Basis of a discourse, 174